50 Pre-Algebra Activities

by

Mary Lou Witherspoon

and

Ernest Woodward

J. WESTON

WALCH

PUBLISHER

Portland, Maine

User's Guide
to
Walch Reproducible Books

As part of our general effort to provide educational materials which are as practical and economical as possible, we have designated this publication a "reproducible book." The designation means that purchase of the book includes purchase of the right to limited reproduction of all pages on which this symbol appears:

Here is the basic Walch policy: We grant to individual purchasers of this book the right to make sufficient copies of reproducible pages for use by all students of a single teacher. This permission is limited to a single teacher, and does not apply to entire schools or school systems, so institutions purchasing the book should pass the permission on to a single teacher. Copying of the book or its parts for resale is prohibited.

Any questions regarding this policy or requests to purchase further reproduction rights should be addressed to:

Permissions Editor
J. Weston Walch, Publisher
321 Valley Street • P. O. Box 658
Portland, Maine 04104-0658

1   2   3   4   5   6   7   8   9   10

ISBN 0-8251-3730-6

Contents

Introduction

The lessons in this book were written to help students develop their algebraic thinking. As students complete these lessons, they should understand the meaning behind algebraic expressions and equations. Although means of solving equations will be explored, the emphasis will be on why they work rather than on a rote memorization of procedures. The major topics addressed are

1. algebraic representations of numerical relationships

2. the concept of equality

3. simple algebraic equations and their solutions

4. graphing numerical and algebraic relationships

5. the concept of functions

This approach is consistent with the National Council of Teachers of Mathematics position concerning the development of algebraic concepts and processes. The following quotes are from page 102 of the well-known NCTM publication *Curriculum and Evaluation Standards for School Mathematics.*

It is essential that students explore algebraic concepts in an informal way to build a foundation for the subsequent formal study of algebra. Such informal explorations should emphasize physical models, data, graphs, and other mathematical representations rather than facility with formal algebraic manipulation.

Learning to recognize patterns and regularities in mathematics and make generalizations about them requires practice and experience. Expanding the amount of time that students have to make this transition to more abstract ways of thinking increases their chances of success. By integrating informal algebraic experiences throughout the K–8 curriculum, students will develop confidence in using algebra to represent and solve problems.

Using This Book

The lessons in this book were written to be used as a supplement to a textbook for a pre-algebra course. In some instances, lessons could be used as a supplement to a standard algebra textbook.

The authors have tried to make the lessons easy to use. Many of the lessons have prerequisite lessons. When that is the case, the prerequisite lessons are listed. Most of the lessons involve worksheets and transparencies. Worksheet masters and transparency masters are provided. Worksheets and transparencies are keyed to individual lessons. For example, the transparency for Lesson 1-1 is labeled "Transparency 1-1," while the two worksheets for this lesson are labeled "Worksheet 1-1-1" and "Worksheet 1-1-2." On the other hand, Lesson 1-2 has only one worksheet, and it is labeled "Worksheet 1-2."

In most cases, it is suggested that lessons be introduced in a large group format and that the worksheets be completed by individual students. It is also possible to use these lessons with individual students or with small groups. Once a lesson is introduced in a large group situation, students can be assigned to groups of three or four, and the worksheets can be completed by the groups.

In most instances, the numbers involved in individual lessons are whole numbers. A special note is included in lessons where computation involves negative integers. It is assumed that students are aware of the conventional order of operations agreements, including the use of grouping symbols and the distributive property.

Chapter 1

Using Variables to Describe Patterns

Students should become comfortable with the use of variables before they take a formal algebra course. This chapter is written with this in mind. Students are presented with situations in which patterns are introduced and are asked to describe the patterns using variables and equations. In Lessons 1-1, 1-2, and 1-3, the numerical patterns are motivated from a geometric perspective. Lesson 1-4 presents students with a situation in which they can explain number tricks by the use of variables. In Lessons 1-5 and 1-6, the data come from a real-world perspective. In Lesson 1-7 and 1-8, the patterns are strictly numerical.

Lesson 1-1: Geometric Patterns I

Objective

To explore geometric relationships and describe results using algebraic notation.

Prerequisite Lessons

None

Materials Needed

For each student: Worksheets 1-1-1, 1-1-2
Transparency 1-1

Directions for the Teacher

In this short lesson, students are to investigate patterns in a geometric context. Place Transparency 1-1 on the overhead projector. Have students examine the patterns. They should conclude that the first array of materials forms one square, the second array forms two squares, and the third array forms three squares. Note that the transparency is actually the first part of the worksheet.

Distribute worksheets. Have students proceed to problem 1 and then to problem 2. Provide individual assistance as needed.

Answer Key

1. (a)

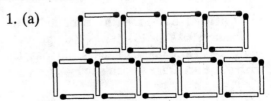

(b)

Squares	Matchsticks
1	4
2	7
3	10
4	13
5	16
6	19
10	31
100	301
n	$3n + 1$ or $4n - (n - 1)$ or $2n + (n + 1)$

2. (a)

(b)

Triangles	Matchsticks
1	3
2	5
3	7
4	9
5	11
6	13
10	21
100	201
n	$2n + 1$ or $3n - (n - 1)$ or $n + (n + 1)$

You might need to give some hints for finding the number of matchsticks for 10 squares, 100 squares, and n squares, and also for 10 triangles, 100 triangles, and n triangles. Encourage students to look for

patterns. Ask students who have difficulty generating an expression involving n to describe in words how they found the number of matchsticks for 100 squares and for 100 triangles. Then help them translate the words to symbols. Note that for the last entry in the table of problem 1b, three possible answers are given. Most students will probably answer $3n + 1$, but the other answers are possible and correct. The same is true for problem 2b.

Lesson 1-2: Geometric Patterns II

Objective

To explore geometric relationships and describe the results using algebraic notation.

Prerequisite Lessons

Lesson 1-1

Materials Needed

For each student: Worksheet 1-2
Transparency 1-2

Directions for the Teacher

Place the transparency on the overhead projector and discuss square numbers. With help from students, draw an array for the fourth square number, then complete the table at the bottom of the page.

Distribute worksheet and have students proceed as directed on the handout.

Answer Key

1. . .   .   .   .

.   .   .   .

.   .   .   .

.   .   .   .

20

2.

RECTANGULAR NUMBERS								
1st	**2nd**	**3rd**	**4th**	**5th**	**6th**	**10th**	**100th**	***n*th**
2	6	12	20	30	42	110	10100	$n(n+1)$

Another possible last entry for the table for problem 2 is $n^2 + n$.

Lesson 1-3: Perimeter Patterns

Objective

To investigate the perimeter of certain polygons and to describe these perimeters with algebraic notation.

Prerequisite Lessons

Lessons 1-1 and 1-2

Materials Needed

For each student: Worksheets 1-3-1, 1-3-2, 1-3-3
One copy of Worksheet 1-3-4 for students who show enthusiasm and want to go further

Directions for the Teacher

Distribute Worksheets 1-3-1, 1-3-2, and 1-3-3. Tell students that you will be working problem 1 on Worksheet 1-3-1 in an entire-class situation. Proceed to the 1-triangle train.

Tell students that each side of this triangle has a length of 1 unit and ask them what the perimeter is (3 units). Proceed to the 2-triangle train. With student help, conclude that the perimeter is 4 units. Similarly conclude that for the 3-triangle train the perimeter is 5 units and for the 4-triangle train the perimeter is 6 units. Have students record this information in their table for problem 1. Then suggest that they look for a pattern concerning the first four entries in their table. With their help, conclude that for a 5-triangle train, the perimeter is 7; for a 10-triangle train, the perimeter is 12; for a 100-triangle train, the perimeter is 102; and for an n-triangle train, the perimeter is n + 2.

Direct students' attention to problem 2. Indicate that in this problem they will be examining square trains, and point out that each side of the square has a length of 1 unit. Briefly comment on problems 3 and 4. Indicate that in each case the polygons have all sides the same length, and each side has a length of 1 unit.

Answer Key

1. Polygon: <u>Triangle</u>
   Number of sides of polygon: <u>3</u>

No. of Polygons	1	2	3	4	5	...	10	100	*n*
Perimeter	3	4	5	6	7	...	12	102	$n+2$

2. Polygon: <u>Square</u>
   Number of sides of polygon: <u>4</u>

No. of Polygons	1	2	3	4	5	...	10	100	*n*
Perimeter	4	6	8	10	12	...	22	202	$2n+2$

3. Polygon: <u>Pentagon</u>
   Number of sides of polygon: <u>5</u>

No. of Polygons	1	2	3	4	5	...	10	100	*n*
Perimeter	5	8	11	14	17	...	32	302	$3n+2$

4. Polygon: <u>Hexagon</u>
   Number of sides of polygon: <u>6</u>

No. of Polygons	1	2	3	4	5	...	10	100	*n*
Perimeter	6	10	14	18	22	...	42	402	$4n+2$

Note that students may arrive at other "correct" answers for the last entries in the tables for problems 1 through 4. It is recommended that Worksheet 1-3-4 be used only with enthusiastic students who exhibit good critical thinking skills. Point out that the information for the first four entries in the table for problem 5 can be found in work for the previous four problems. Suggest that students should look for patterns. This should help them with the last three entries in this table.

5.

No. of sides of the polygon	Perimeter of a train with *n* polygons
3	$n + 2$
4	$2n + 2$
5	$3n + 2$
6	$4n + 2$
7	$5n + 2$
10	$8n + 2$
.	.
.	.
.	.
m	$(m - 2)\, n + 2$

Lesson 1-4: Number Tricks

Objective

To analyze certain number tricks using algebra.

Prerequisite Lessons

None

Materials Needed

For each student: Worksheet 1-4
Transparency 1-4

Directions for the Teacher

Place the transparency on the overhead projector and go through the directions given there. Select one student to come to the overhead projector. Tell the student to select a number and then do the computation described, writing the result for each step in the blank provided. Tell students that you will be able to determine the original number without looking at the student's selection. Walk away from the overhead projector, face away from the screen, and have the student follow the directions given. When the student

has finished, ask him or her what the final number is. Divide that number by 2; this will be the original number. Repeat the procedure with another student. Then ask the students how they think the trick works.

Suggest that you call the original number *x*. Then do the algebraic computation, step by step, and write these results beside the directions.

1. Select an integer greater than 3. x
2. Subtract 3. $x - 3$
3. Multiply by 2. $2x - 6$
4. Add 6. $2x$

Conclude that in each case, the resulting number is 2 times the original number, so you can find the original number by dividing the final number by 2.

Distribute a copy of the worksheet to each student. Break the class up into groups of three or four. Have the groups investigate the tricks. A general solution for each problem is given below.

Answer Key

1. Select an integer. x

   Multiply by 2. $2x$

   Add 1. $2x + 1$

   Multiply by 5. $10x + 5$

   Subtract 5. $10x$

   (a) The final number was 70. What was the original number? 7

   (b) The final number was 120. What was the original number? 12

   (c) The final number was 24,490. What was the original number? $2,449$

2. Select an even positive integer. x

   Divide by 2. $\dfrac{x}{2}$

   Add 3. $\dfrac{x}{2} + 3$

   Multiply by 2. $x + 6$

   Subtract 5. $x + 1$

(a) The final number was 15. What was the original number?   14

(b) The final number was 27. What was the original number?   26

(c) The final number was 24,755. What was the original number?   24,754

Lesson 1-5: Number Relationship Applications

Objective

To model number relationships using equations.

Prerequisite Lessons

Lessons 1-1 and 1-2.

Materials Needed

For each student: Worksheets 1-5-1, 1-5-2

Directions for the Teacher

Distribute Worksheets. Have the students proceed to problem 1. Emphasize that they are to complete the table but they are **not** to introduce new letters. Comparable comments are appropriate for problem 3. Provide individual assistance as needed.

Answer Key

1.

No. of Girls in Class	No. of Boys in Class	No. of Students in Class
17	15	32
13	16	29
10	16	26
g	11	$g + 11$
$25 - b$	b	25

14	$s - 14$	s
$s - 17$	17	s
g	$s - g$	s
$s - b$	b	s

2. $g + b = s$ Two of those three are
   $s - g = b$ appropriate for parts a and b.
   $s - b = g$

3.

No. of Pieces of Candy in Each Container	No. of Containers	Total No. of Pieces of Candy
6	4	24
5	7	35
8	6	48
6	c	$6c$
p	7	$7p$
$t \div 5$	5	t
11	$t \div 11$	t
$t \div c$	c	t
p	$t \div p$	t

4. $pc = t$ Two of these three are
   $t \div p = c$ appropriate for parts a and b.
   $t \div c = p$

Lesson 1-6: Postal Rate Patterns

Objective

To examine postal rate patterns and to describe these relationships with equations.

Prerequisite Lessons

Lessons 1-1, 1-2, 1-4, and 1-5.

Materials Needed

For each student: Worksheet 1-6
Transparency 1-6

Directions for the Teacher

With student help complete the table.

Weight in Ounces	Cost in Cents
5	<u>105</u>
10	<u>205</u>
20	<u>405</u>
<u>26</u>	525
w	<u>$25 + (w{-}1)20$</u> or <u>$20w + 5$</u>

Students may easily find the entry corresponding to 5 ounces (add 20 to the previous entry), but might have difficulty with other entries. Encourage them to look at patterns. Notice that there are two reasonable entries corresponding to the weight entry of w. Some students may notice that it costs 25¢ for the first ounce and 20¢ for each additional ounce, which makes the $25 + (w{-}1)20$ entry reasonable. Others may notice that in each case, the "cost in cents" entry is 5 more than 20 times the number of ounces. This makes the $20w + 5$ entry reasonable. If both entries come up in the discussion, indicate that

$$25 + (w{-}1)20 = 20w + 5$$

Suggest that it would be reasonable to allow c to represent the cost of mailing a letter of w ounces. Indicate then that $c = 20w + 5$, or $c = 25 + (w{-}1)20$.

Distribute worksheets and provide individual assistance as needed.

Answer Key

1. (a)

Weight in Ounces	Cost in Cents
5	<u>121</u>
11	<u>259</u>
<u>20</u>	466
w	<u>$29 + 23(w{-}1)$</u> or <u>$23w + 6$</u>

(b) $c = 29 + 23(w{-}1)$ or $c = 23w + 6$

2 (a)

Weight in Ounces	Cost in Cents
5	<u>57</u>
12	<u>134</u>
<u>21</u>	233
w	<u>$13 + 11(w{-}1)$</u> or <u>$11w + 2$</u>

(b) $c = 11w + 2$ or $c = 13 + 11(w{-}1)$

Lesson 1-7: Numerical Patterns I

Objective

To analyze tables, identify relationships, and describe relationships algebraically.

Prerequisite Lessons

Lessons 1-1, 1-2, 1-4, and 1-5

Materials Needed

For each student: Worksheet 1-7-1

One copy of Worksheet 1-7-2 for students who can perform computation in integers

Transparency 1-7

Directions for the Teacher

Place the transparency on the overhead projector and proceed to problem 1. Ask students to look for a pattern involving the relationship between the first number and the second number. They will probably see that you can get the second number by multiplying the first number by 5. With students, complete this table.

First Number	Second Number
100	<u>500</u>
<u>12</u>	60
n	<u>5n</u>
<u>4x</u>	20x

Direct students to look for a pattern in problem 2. Help them conclude that you arrive at the second number by subtracting 3 from the first number or by adding –3 to the first number. With their help, complete this table.

First Number	Second Number
21	<u>18</u>
37	<u>34</u>
<u>48</u>	45
<u>81</u>	78
x	<u>x – 3 or x + –3</u>
<u>y + 3</u>	y

Distribute a copy of the first worksheet. Have students proceed to problem 1 and then to problem 2. Provide individual assistance as needed. If students have difficulty with problem 2, suggest that they initially cover up the third column and work with the first two columns. Then they could cover up the second column and work with columns one and three.

Answer Key

1.

First Number	Second Number
4	<u>9</u>
<u>21</u>	26
n	<u>n + 5</u>
<u>x</u>	x + 5
y + 2	<u>y + 7</u>
<u>z + 8</u>	z + 13

2.

First Number	Second Number	Third Number
9	<u>18</u>	<u>54</u>
8	<u>16</u>	<u>48</u>
<u>6</u>	12	<u>36</u>
<u>12</u>	<u>24</u>	72
<u>4</u>	<u>8</u>	24
x	<u>2x</u>	<u>6x</u>
<u>2y</u>	4y	<u>12y</u>
<u>3z</u>	<u>6z</u>	18z

Distribute a copy of the second worksheet page to those students who are able to perform computation involving integers (including negative integers). Again, provide individual assistance as needed.

3.

First Number	Second Number
11	<u>23</u>
–15	<u>–29</u>
<u>23</u>	47
<u>–31</u>	–61
x	<u>2x + 1</u>
<u>y</u>	2y + 1
3z	<u>6z + 1</u>
<u>4w</u>	8w + 1

Lesson 1-8: Numerical Patterns II

Objective

To analyze tables, identify relationships, and describe relationships algebraically.

Prerequisite Lessons

Lessons 1-1, 1-2, 1-4, 1-5, and 1-7

Materials Needed

For each student: Worksheet 1-8
Transparency 1-8

Directions for the Teacher

Place the transparency on an overhead projector. Direct students' attention to the first and second columns, asking them if they can see a pattern and, more specifically, how they could get the second number from the original number. Complete the next three rows for the first two columns.

Next, have them investigate the relationship between the numbers in the first column and those in the third column. Emphasize that they are to disregard (temporarily) the second column. They will probably need a few hints. If they don't initially see a pattern, suggest multiplying the numbers in the first column by 3. As a class, complete the table.

Original Number	Second Number	Third Number
7	14	22
1	2	4
6	12	19
4	8	13
x	$2x$	$3x + 1$

Distribute worksheet.

Answer Key

Original Number	Second Number	Third Number	Fourth Number
12	25	9	144
10	21	7	100
7	15	4	49
9	19	6	81
20	41	17	400
x	$2x + 1$	$x - 3$	x^2

Squares can be made using matchsticks. The squares in this lesson will be made in a special way. This is shown below.

First

Second

Third

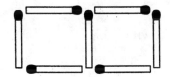

50 Pre-Algebra Activities

Geometric Patterns I

1. Squares can be made using matchsticks. The squares in this lesson will be made in a special way. This is shown below.

First   **Second**   **Third**

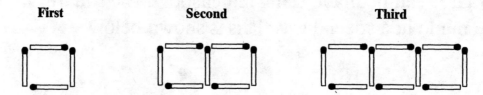

a. Draw pictures of the next two arrays of squares.

b. Complete the following table.

Number of Squares	Number of Matchsticks
1	
2	
3	
4	
5	
6	
10	
100	
n	

Geometric Patterns I

2. Triangles can be made using matchsticks. The triangles in this lesson will be made in a special way. This is shown below.

First   **Second**   **Third**

a. Draw pictures of the next two arrays of triangles.

b. Complete the following table.

Number of Triangles	Number of Matchsticks
1	
2	
3	
4	
5	
6	
10	
100	
n	

1. Numbers that can be represented by a square array of dots are called **square numbers**. The first three square numbers, together with the arrays that go with them, are given below.

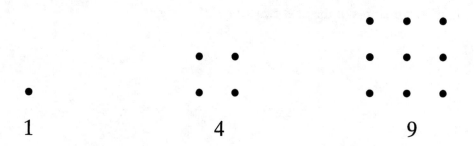

1   4   9

Make an array for the next square number.

2. Complete the table below.

SQUARE NUMBERS								
1st	2nd	3rd	4th	5th	6th	10th	100th	*n*th

▬▬▬▬▬▬ Geometric Patterns II ▬▬▬▬

1. Numbers that can be pictured as a rectangular array of dots are called rectangular numbers, if there is one more dot in each horizontal row than in each vertical column. The first three rectangular numbers and the arrays that go with them are pictured below.

```
•   •              •   •   •              •   •   •   •

   2               •   •   •              •   •   •   •
                       6
                                          •   •   •   •
                                              12
```

Draw an array for the next rectangular number.

2. Complete the table below.

RECTANGULAR NUMBERS								
1st	**2nd**	**3rd**	**4th**	**5th**	**6th**	**10th**	**100th**	*n*th

Name _____

Date _____

■ **Perimeter Patterns** ■

1. Study the trains below and complete the table.

One-Triangle Train

Two-Triangle Train

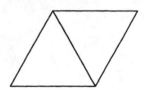

Three-Triangle Train

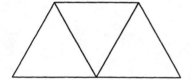

Four-Triangle Train

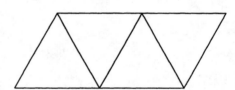

Polygon:_____   Number of sides of polygon:_____

Number of Polygons	1	2	3	4	5	. . .	10	100	n
Perimeter									

Perimeter Patterns

2. Study the trains below and complete the table.

One-Square Train

Two-Square Train

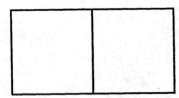

Three-Square Train

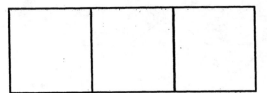

Four-Square Train

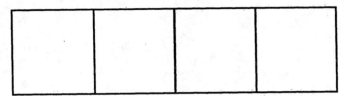

Polygon:_____   Number of sides of polygon:_____

Number of Polygons	1	2	3	4	5	...	10	100	n
Perimeter									

Perimeter Patterns

3. Study the trains below and complete the table.

One-Pentagon Train

Two-Pentagon Train

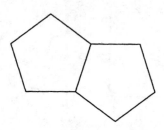

Three-Pentagon Train

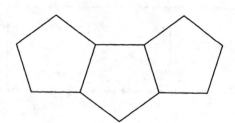

Four-Pentagon Train

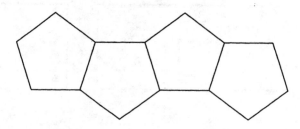

Polygon:_____ Number of sides of polygon:_____

Number of Polygons	1	2	3	4	5	...	10	100	n
Perimeter									

4. Study the tables for problems 1, 2, and 3 and look for patterns. Then complete the table for hexagons.

Polygon: <u>Hexagon</u> Number of sides of polygon: <u>6</u>

Number of Polygons	1	2	3	4	5	...	10	100	n
Perimeter									

Perimeter Patterns

5. Complete the following table.

Number of sides of the polygon	Perimeter of a train with n polygons
3	
4	
5	
6	
7	
10	
. . .	
m	

1. Select an integer greater than 3. _____

2. Subtract 3. _____

3. Multiply by 2. _____

4. Add 6. _____

Number Tricks

The two number trick problems below are similar to the one discussed in class. In each case, if a person selects a number and follows the directions, you should be able to tell him or her what the original number was. Study each trick and determine what the original number was with given final numbers.

1. Select an integer.   _____

   Multiply by 2.   _____

   Add 1.   _____

   Multiply by 5.   _____

   Subtract 5.   _____

   (a) The final number was 70.
       What was the original number?   _____

   (b) The final number was 120.
       What was the original number?   _____

   (c) The final number was 24,490.
       What was the original number?   _____

2. Select an even positive integer.   _____

   Divide by 2.   _____

   Add 3.   _____

   Multiply by 2.   _____

   Subtract 5.   _____

   (a) The final number was 15.
       What was the original number?   _____

   (b) The final number was 27.
       What was the original number?   _____

   (c) The final number was 24,755.
       What was the original number?   _____

19   *50 Pre-Algebra Activities*

Number Relationship Applications

1. The partially completed table below shows the number of girls in class, the number of boys in class, and the total number of students in class. Complete the table. Do *not* introduce new letters (variables).

Number of Girls in Class	Number of Boys in Class	Number of Students in Class
17	15	
13		29
	16	26
g	11	
	b	25
14		s
	17	s
g		s
	b	s

2. Suppose that *g* represents the number of girls in class, *b* represents the number of boys in class and *s* represents the number of students in class.

(a) Write one equation that illustrates the relationship among *g*, *b*, and *s*.

(b) Write a different equation that illustrates the relationship among *g*, *b*, and *s*.

Number Relationship Applications

3. The partially completed table below shows the number of pieces of candy in each container, the number of containers, and the total number of pieces of candy. Complete this table. Do *not* introduce new letters (variables).

Number of Pieces of Candy in Each Container	Number of Containers	Total Number of Pieces of Candy
6	4	
	7	35
8		48
6	c	
p	7	
	5	t
11		t
	c	t
p		t

4. Suppose *p* represents the number of pieces of candy in each container, *c* represents the number of containers, and *t* represents the total number of pieces of candy.

(a) Write one equation that illustrates the relationship among *p*, *c*, and *t*.

(b) Write another equation that illustrates the relationship among *p*, *c*, and *t*.

1. (a) The partially completed table below shows first-class postal rates for the years immediately prior to 1994. Complete this table.

Weight in Ounces	Cost in Cents
1	25
2	45
3	65
4	85
5	
10	
20	
	525
w	

(b) Suppose c is used to represent the cost of mailing a package that weighs w ounces. Write an equation that describes this cost.

Postal Rate Patterns

1. (a) The partially completed table below shows 1995 first-class postal rates. Complete the table.

Weight in Ounces	Cost in Cents
1	29
2	52
3	75
4	98
5	
11	
	466
w	

   (b) Suppose c is used to represent the cost of mailing a package that weighs w ounces. Write an equation that describes this cost.

2. (a) The partially completed table below shows first-class postal rates for the mid 1970's. Complete the table.

Weight in Ounces	Cost in Cents
1	13
2	24
3	35
4	46
5	
12	
	233
w	

   (b) Suppose c is used to represent the cost of mailing a package that weighs w ounces. Write an equation that describes this cost.

23   *50 Pre-Algebra Activities*

Using the pattern shown in the first few lines, complete these tables.

1.

First Number	Second Number
3	15
7	35
10	50
21	105
100	
	60
n	
	20x

2.

First Number	Second Number
3	0
5	2
9	6
14	11
21	
37	
	45
	78
x	
	y

50 Pre-Algebra Activities

Numerical Patterns I

Using the pattern shown in the first few lines, complete the tables which follow.

1.

First Number	Second Number
3	8
5	10
7	12
10	15
4	
	26
n	
	$x + 5$
$y + 2$	
	$z + 13$

2.

First Number	Second Number	Third Number
2	4	12
5	10	30
8	16	48
1,000	2,000	6,000
10	20	60
9		
8		
	12	
		72
		24
x		
	$4y$	
		$18z$

50 Pre-Algebra Activities

Numerical Patterns I

3.

First Number	Second Number
1	3
3	7
5	11
8	17
−5	−9
11	
−15	
	47
	−61
x	
	$2y + 1$
$3z$	
	$8w + 1$

Complete the table given below.

Original Number	Second Number	Third Number
2	4	7
5	10	16
3	6	10
7		
1		
	12	
		13
x		

50 Pre-Algebra Activities

Numerical Patterns II

1. Study the completed portion of the table given below. Look for patterns and then complete the table.

Original Number	Second Number	Third Number	Fourth Number
3	7	0	9
5	11	2	25
8	17	5	64
6	13	3	36
12			
10			
		4	
			81
	41		
x			

50 Pre-Algebra Activities

Chapter 2

Introduction to Equality and Equations

The purpose of this chapter is to help students find ways to solve simple equations. The equations investigated in this chapter are basically of this type:

$$ax + b = c \qquad\qquad (x + a) \bullet b = c$$

$$ax - b = c \qquad\qquad (x - a) \bullet b = c$$

$$x \div a + b = c \qquad\qquad (x + a) \div b = c$$

$$x \div a - b = c \qquad\qquad (x - a) \div b = c$$

In Lesson 2-1, students learn about equality. In Lesson 2-2, they learn how equality and equations are related and what it means to have a solution for an equation. In Lesson 2-3, with arrow pictures, students learn about inverse operations. In Lessons 2-4 and 2-5, they learn about scales. In Lesson 2-6, they learn how scales can be used to develop equivalent equations and equation-solving procedures. In Lessons 2-7 and 2-8, they learn to solve problems and related equations using arrow pictures. Lesson 2-9 consists of word problems that can be solved with algebra.

Lesson 2-1: Equality

Objectives

To recognize that "=" expresses a significant relationship between two quantities

Prerequisite Lessons

None

Materials Needed

For each student: Worksheet 2-1
Transparency 2-1

Directions for the Teacher

Place the transparency on an overhead projector and read the discussion of the meaning "=." Proceed to problems 1 and 2. With students, conclude that $3 + 5 = 4 + 4$ is a true sentence because

$$3 + 5 = 8 \text{ and}$$

$$4 + 4 = 8$$

but $3 \cdot 5 = 10 \div 2 + 8$ is false because

$$3 \cdot 5 = 15 \text{ and}$$

$$10 \div 2 + 8 = 13$$

$x + 3 = 5$ is neither true nor false. It is an open sentence. Substituting 2 for x will make it true. Substituting any number other than 2 for x will make it false.

Distribute worksheet. Provide individual help as needed.

Answer Key

1. (a) True

   (b) False

   (c) True

   (d) False

2. (a) $7 \cdot 4 = 23 + \underline{5}$

   (b) $\underline{3} \cdot 6 = 25 - 7$

   (c) $\underline{12} \div 3 = 8 - 4$

   (d) $2 + 2 = 36 \div \underline{9}$

Lesson 2-2: Introduction to Equations

Objective

To determine whether particular numbers are solutions to given equations and to write equations with given numbers as solutions.

Prerequisite Lessons

Lesson 2-1

Materials Needed

For each student: Worksheet 2-2-1

One copy of Worksheet 2-2-2 for students who are comfortable with computation involving integers

Transparency 2-2

Directions for the Teacher

Place the transparency on an overhead projector and present the information at the top of the page. Proceed to problem 1. In the equation $2x - 1 = 7$, replace x by 1, 2, 3, and 4 and you will get the sentences:

$$2 \cdot 1 - 1 = 7$$

$$2 \cdot 2 - 1 = 7$$

$$2 \cdot 3 - 1 = 7$$

$$2 \cdot 4 - 1 = 7$$

Conclude that the only true statement is the last one and that 4 is the solution. Then write $x = 4$.

Proceed to problem 2. Point out that this equation is different from problem 1 in that x occurs twice. Emphasize that in every instance you must replace each x by the selected number. When this is done for the specified numbers, the following sentences result:

$$(4 - 2)(4 + 3) = 36$$

$$(5 - 2)(5 + 3) = 36$$

$$(6 - 2)(6 + 3) = 36$$

$$(7 - 2)(7 + 3) = 36$$

Conclude that the only true statement is the one where x was replaced by 6 and that 6 is a solution. Then write $x = 6$.

Let individual students or small groups of students work on problem 3. Write all these student-generated equations on the blackboard, and check to see that $x = 5$ is actually a solution of each given equation. There are many possible answers for this problem, but $2x = 10$, $x - 5 = 0$ and $x + 1 = 6$ are possibilities.

Distribute Worksheet 2-2-1. Provide individual help as needed. Distribute a copy of Worksheet 2-2-2 to appropriate students.

Answer Key

1. $x = 6$	4. $x = 2$
2. $x = 7$	5. $x = 3$
3. $x = 3$	6. $x = 6$

˙Answers to problems 6 through 9 are not unique. Possible correct answers follow.

7. $2x = 8$

8. $x + 3 = 5$

9. $(x - 1)(x - 3) = 0$

Students might find problem 9 difficult. Some hints may be needed.

10. $x = -5$

11. $x = -6$

12. $x = -5$ and $x = -7$

Students may be surprised that both –5 and –7 are solutions of the equation given in problem 12.

Lesson 2-3: Arrow Pictures and Inverse Operations

Objective

To learn that addition and subtraction are inverse operations and multiplication and division are inverse operations.

Prerequisite Lessons

None

Materials Needed

For each student: Worksheets 2-3-1, 2-3-2
Transparencies 2-3-1, 2-3-2

Directions for the Teacher

Place Transparency 2-3-1 on an overhead projector and proceed to problem 1. With students, label the dots. Proceed to problem 2. Again, with students, label the dots. Correct answers follow.

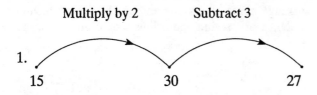

Subtract 10   Multiply by 3   Divide by 2

2.
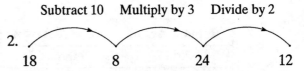
18   8   24   12

Students may have a little trouble labeling the dot on the left for problem 2. If so, suggest that they are looking for a number such that when you subtract 10 from that number, the result is 8.

Place Transparency 2-3-2 on an overhead projector and proceed to problem 1. With students, label each point. Draw in the reverse arrows, and ask students what each reverse arrow represents (subtract 3). Go on to problem 2. Again, label each point and draw in all the reverse arrows and ask students what each reverse arrow represents (divide by 2).

Add 3   Add 3   Add 3

1.
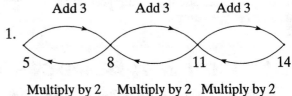
5   8   11   14

Multiply by 2   Multiply by 2   Multiply by 2

2.

3   6   12   24

Distribute worksheets and provide individual help as needed.

Answer Key

1.   Subtract 10   Subtract 10   Subtract 10

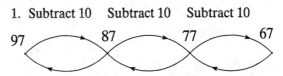
97   87   77   67

(c) Each reverse arrow represents adding 10.

2.   Divide by 3   Divide by 3   Divide by 3

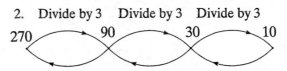
270   90   30   10

(c) Each reverse arrow represents multiplying by 3.

3.

Regular Arrow	Reverse Arrow
Add 2	Subtract 2
Subtract 5	Add 5
Divide by 7	Multiply by 7
Multiply by 100	Divide by 100
Multiply by 3	Divide by 3
Subtract 7	Add 7

After students finish the worksheets, ask them to investigate the four standard operations and how these operations are related. They should conclude that subtraction and addition undo each other. Point out that they are called inverse operations. They should also conclude that multiplication and division are inverse operations.

Lesson 2-4: Balance-Preserving Operations

Objective

To explore the concept of equality in the context of balance scales.

To discover balance-preserving actions.

Prerequisite Lessons

Lessons 2-1 and 2-2

Materials Needed

For each student: Worksheets 2-4-1, 2-4-2
Transparencies 2-4-1, 2-4-2

Directions for the Teacher

Place transparency 2-4-1 on the overhead projector. Point out that figures that look alike have the same weight and that figures that look different probably have different weights.

Ask students to compare scales 1 and 2. If scale 1 balances, why should scale 2 balance? Students should note that the same amount of weight has been subtracted from both sides. Emphasize that subtracting the same weight from both sides of the scale is a balance-preserving operation.

Ask students to compare scales 1 and 3. If scale 1 balances, why should scale 3 balance? Students should note that the same amount of weight has been added to both sides. Emphasize that adding the same weight to both sides of the scale is a balance-preserving operation.

Ask students to compare scales 1 and 4. If scale 1 balances, why should scale 4 balance? Students should note that the weight has been doubled on both sides. Emphasize that multiplying the weight on both sides of the scale by the same amount is a balance-preserving operation.

Ask students to compare scales 1 and 5. If scale 1 balances, why should scale 5 balance? Students should note that the weight has been divided by 2 on both sides. Emphasize that dividing the weight on both sides of the scale by the same amount is a balance-preserving operation.

Enrichment question: What are some possible weights for the square and the circle?

Solution: Many different amounts are possible. Students should notice that a square weighs as much as two circles. If the circle weighs 2 grams, the square weighs 4 grams, and so on.

Distribute worksheets. Students should use balance-preserving operations on the original scale in each problem to generate new arrangements of the weights that balance. In each case, they should explain why the new arrangement balances. Discuss students' responses, pointing out as many different arrangements as possible and emphasizing which balance-preserving operations they used.

Answer Key

1. (a),(b) There are many possible solutions. A few are shown here:

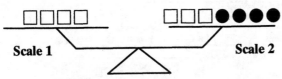

Scale 1   **Scale 2**

Remove ● from each side of Scale 2.

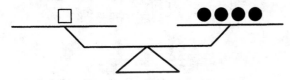

Remove ● ☐☐☐ from each side of Scale 2.

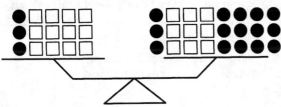

Triple the weight on each side.

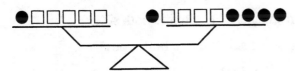

Add ☐ to each side of Scale 2.

1. (c) (Other solutions are possible.)

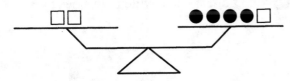

Remove ☐☐ ● from each side of Scale 2.

2. (a),(b) There are many possible solutions. A few are shown here:

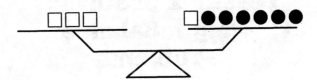

● was removed from each side of Scale 2.

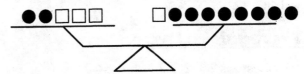

● was added to each side of Scale 2.

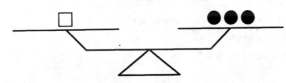

☐ ● was removed from each side. The weight on both sides was then divided by 2.

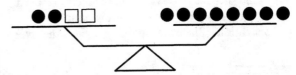

● was added to each side and ☐ was removed from each side.

2. (c)

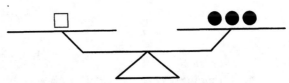

Remove ● from each side of Scale 2. Then remove ☐ from each side of the scale. Divide the weight on each side by 2.

Place Transparency 2-4-2 on an overhead projector. The left column contains a summary of the balance-preserving operation. Note that each balance-preserving operation has a corresponding equality-preserving operation.

Lesson 2-5: Solving Weight Balance Problems

Objective

To use the balance scale to find the unknown weight

Prerequisite Lessons

Lessons 2-1, 2-2, 2-3, and 2-4

Materials Needed

For each student: Worksheet 2-5
Transparency 2-4-2
Transparency 2-5

Directions for the Teacher:

Place Transparency 2-4-2 from lesson 2-4 on an overhead projector. Review the balance-preserving operations and their corresponding equality-preserving operations. Place transparency 2-5 on the over-head projector. Show the problems one at a time. Let students think about them. Have students share their solutions. Emphasize which balance-preserving operations were used.

Solutions:

1. ☐ weighs 9 grams. Subtract 2 grams from each side. ☐ ☐ weighs 18 grams and 18 ÷ 2 = 9, so ☐ weighs 9 grams.

2. ☐ weighs 4 grams. Subtract ☐ ☐ ☐ from both sides. Divide weight on both sides by 3. Note that the grams can be split into 3 groups of 4.

Distribute Worksheet 2-5. Emphasize that students write out their explanations.

Answer Key

1. ☐ weighs 6 grams.   Remove 3 ● from each side. Divide both sides by 2.

2. ☐ weighs 2 grams.   Remove ☐ from each side. Divide both sides by 5.

3. ☐ weighs 3 grams.   Option 1: Divide both sides by 2.
   Remove ☐●●●●● from each side.

   Option 2: Remove ☐☐●●●●●●●●●● from both sides. Divide both sides by 2.

4. ☐ weighs 5 grams.   Remove ☐●● from both sides. Divide both sides by 3.

Lesson 2-6: Linking Scales and Equations

Objective

To link a balance scale situation to an equation.

To link operations that preserve balance to mathematics operations that preserve equality in order to solve equations.

Prerequisite Lessons

Lessons 2-1, 2-2, 2-3, 2-4, and 2-5

Materials Needed

For each student: Worksheets 2-6-1, 2-6-2, 2-6-3, 2-6-4, 2-6-5

Transparency 2-4-2

Transparencies 2-6-1, 2-6-2, 2-6-3

Directions for the Teacher

Place Transparency 2-4-2 on the overhead projector. Review balance-preserving operations and the corresponding equality-preserving operations. Place Transparency 2-6-1 on the overhead projector. Cover

the "reasons" column. Discuss why each equation relates to the situation. Emphasize that finding a solution is getting a value for x that makes the equation a true statement. After the discussion, uncover the "reasons" column. Follow up by checking to see whether the scale would balance if ☐ weighed 1 gram or 3 grams. Would the equation be true if $x = 1$ or $x = 3$?

Place Transparency 2-6-2 on the overhead projector. Cover the "reasons" column and the second scale. Have students supply the first equation and the reason for the first equation, $3x = 12$. Reason: x represents the weight of ☐. There are 3 ☐ on the left side and there are 12 grams on the right side. In a class discussion, have students find the weight of ☐. After the discussion, uncover the "reasons" column and the 2nd scale. The final equation is $x = 4$.

Place Transparency 2-6-3 on the overhead projector. Possibly give students a few minutes to think about how to find the weight of ☐. Ask students how to generate an equation to represent the balanced scale. They should explain why their equation represents the balanced scale. Ask students to think about what they could do to the scales that would preserve balance and help them find the weight of ☐. Record the scales, equations and equality-preserving operations on the overhead. (Refer to Transparency 2-4-2.)

Scales and equations:

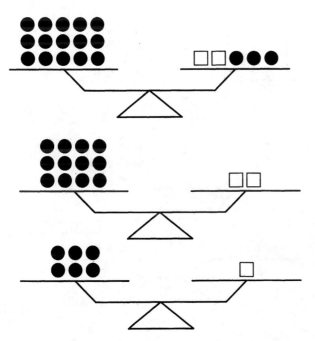

$15 = 2x + 3$

$15 - 3 = (2x + 3) - 3$ EPO – <u>S</u>

Note: $(2x + 3) - 3 = 2x$ because adding 3 and subtracting 3 are inverse operations.

$12 = 2x$

$12 \div 2 = 2x \div 2$ EPO – <u>D</u>

Note: $2x \div 2 = x$ because multiplying by 2 and dividing by 2 are inverse operations.

$6 = x$

Check: $15 = 2 \cdot \underline{6} + 3$ is a true statement.

Follow up by asking students if 5 or 7 is a solution.

Distribute Worksheets 2-6-1, 2-6-2, and 2-6-3 and then discuss as a class whether to work independently or cooperatively.

Answer Key

Solutions to Worksheets 2-6-1, 2-6-2, 2-6-3

1. $14 = x + 8$

   $14 - 8 = (x + 8) - 8$ EPO–$\underline{S}$

   $6 = x$

   Check: $14 = \underline{6} + 8$ is a true statement.

2. $3x = 15$

   $3x \div 3 = 15 \div 3$ EPO–$\underline{D}$

   $x = 5$

   Check: $3 \cdot \underline{5} = 15$ is a true statement.

3. $4x + 5 = 25$

   $(4x + 5) - 5 = 25 - 5$ EPO–$\underline{S}$

   $4x = 20$

   $4x \div 4 = 20 \div 4$ EPO–$\underline{D}$

   $x = 5$

   Check: $4 \cdot \underline{5} + 5 = 25$ is a true statement.

4. $x + 7 = 2x$

   $(x + 7) - x = 2x - x$ EPO–$\underline{S}$

   $7 = x$

   Check: $\underline{7} + 7 = 2 \cdot \underline{7}$ is a true statement.

5. $3x = x + 8$

   $3x - x = (x + 8) - x$ EPO–$\underline{S}$

   $2x = 8$

   $2x \div 2 = 8 \div 2$ EPO=$\underline{D}$

   $x = 4$

   Check: $3 \cdot \underline{4} = \underline{4} + 8$ is a true statement.

6. $4x + 3 = x + 15$

   $(4x + 3) - 3 = (x + 15) - 3$ EPO–$\underline{S}$

   $4x = x + 12$

   $4x - x = (x + 12) - x$ EPO–$\underline{S}$

   $3x = 12$

   $3x \div 3 = 12 \div 3$ EPO–$\underline{D}$

   $x = 4$

   Check: $4 \cdot \underline{4} + 3 = \underline{4} + 15$ is a true statement. (There are other ways to adjust the weights.)

Distribute Worksheets 2-6-4 and 2-6-5 for students to work independently or cooperatively and then discuss as a class.

Solutions to Worksheets 2-6-4, 2-6-5

1.

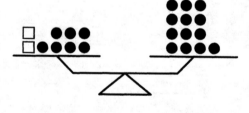

$\boxed{2x = 6}$

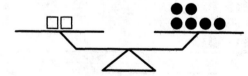

$\boxed{x = 3}$

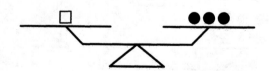

Check: $2 \cdot \underline{3} + 7 = 6 + 7 = 13$

Thus, $2 \cdot \underline{3} + 7 = 13$

2.

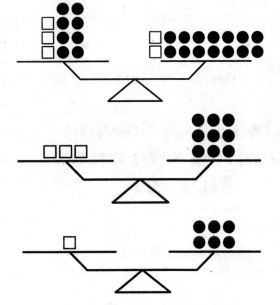

$3x = 2x + 6$

$x = 6$

Check: $3 \cdot \underline{6} + 8 = 18 + 8$ and $2 \cdot \underline{6} + 14 = 12 + 14$
$= 26$ $= 26$

Thus, $3 \cdot \underline{6} + 8 = 2 \cdot \underline{6} + 14$

3.

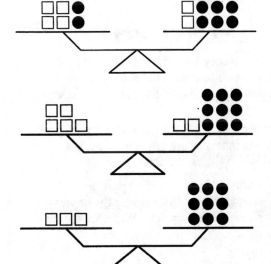

$5x = 2x + 9$

$3x = 9$

$\boxed{x = 3}$

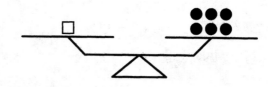

Check: $5 \cdot \underline{3} + 3 = 15 + 3$ and $2 \cdot \underline{3} + 12 = 6 + 12$
$= 18$ $= 18$
Thus, $5 \cdot \underline{3} + 3 = 2 \cdot \underline{3} + 12$

Lesson 2-7: Solving Equations with Arrow Pictures

Objective

To understand the reasons behind methods for solving equations by relating equations to arrow pictures

Prerequisite Lessons

Lessons 2-1, 2-2, 2-3, 2-4, 2-5, and 2-6

Materials Needed

For each student: Worksheets 2-7-1, 2-7-2, 2-7-3, 2-7-4
Transparency 2-4-2
Transparencies 2-7-1, 2-7-2

Directions for the Teacher

Place Transparency 2-4-2 on an overhead projector and review the relationship between balance-preserving operations and equality-preserving operations.

Place Transparency 2-7-1 on an overhead projector. Tell students that they are going to introduce using variables into the use of arrow pictures.

1. (a) Follow the directions to part (a) of the transparency. Discuss why the expressions are appropriate. Write the expres-

sions on the transparency above the dots as shown here.

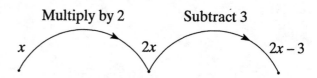

(b) Students should evaluate each expression as they make the indicated substitution for x. Write the appropriate numbers below the dots for the first example as shown here.

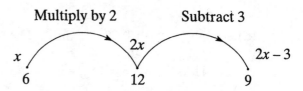

Point out that when $x = 6$, $2x = 12$ and $2x - 3 = 9$. Because all three equations have the same solution, they are said to be equivalent. If we substitute 6 for x in each of them, all three resulting statements will be true. Erase the numbers below the dots but keep the expressions above the dots and work the second example. Again, point out that when $x = 14$, $2x = 28$ and $2x - 3 = 25$. These three equations are equivalent to each other. If we substitute 14 for x in each of them, all three resulting statements will be true.

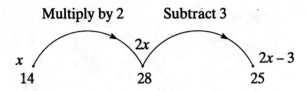

(c) Erase the numbers below the dot. Encourage students to use inverse operations to find the value for *x*. Students should use the inverse operations to "undo" each operation. Record the inverse operation arrows and the resulting values on the arrow picture as shown here.

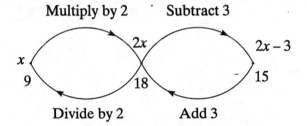

(d) In the boxes, record the equivalent equations suggested by the arrow picture. In the line below each equation show what was done to produce the next equation. Justify the action with one of the equality-preserving operations from Transparency 2-4-2.

Equation $\boxed{2x - 3 = 15}$

$\qquad \underline{(2x - 3) + 3 = 15 + 3}$ EPO–<u>A</u>

Equation $\boxed{2x = 18}$

$\qquad \underline{2x \div 2 = 18 \div 2}$ EPO–<u>D</u>

Equation $\boxed{x = 9}$

Note that the final equation gives the solution to $2x - 3 = 15$. Emphasize that $x = 9$ is a solution, because when 9 is substituted for *x* in the equation, the resulting statement is true.

Place Transparency 2-7-2 on an overhead projector.

2. (a) Students should tell what operations are represented by the arrows based upon the expressions involving *x*.

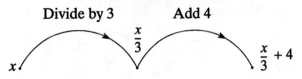

(b) Students should use inverse operations to solve the equation given. During the discussion, record on the overhead the inverse-operation arrows and the resulting values for each expression as shown here.

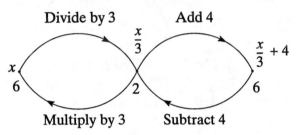

(c) Students should identify the three equations generated by the arrow picture.

$\boxed{\dfrac{x}{3} + 4 = 6}$

$\dfrac{x}{3} + 4 - 4 = 6 - 4$ $\qquad$ EPO–<u>S</u>

$\boxed{\dfrac{x}{3} = 2}$

$\dfrac{x}{3} \cdot 3 = 2 \cdot 3$ $\qquad$ EPO-<u>M</u>

$\boxed{x = 6}$

Emphasize that $x = 6$ is the solution because substituting 6 for *x* makes a true statement.

Give each student one copy of each worksheet page to work either independently or cooperatively.

Answer Key

1. (a)

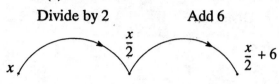

Divide by 2   Add 6

x → $\frac{x}{2}$ → $\frac{x}{2} + 6$

(b)

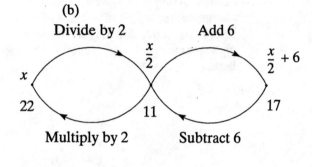

Divide by 2   Add 6

$\frac{x}{2}$ → $\frac{x}{2} + 6$

x 22   11   17

Multiply by 2   Subtract 6

(c) $\boxed{\dfrac{x}{2} + 6 = 17}$

$\dfrac{x}{2} + 6 - 6 = 17 - 6$ EPO–<u>S</u>

$\boxed{\dfrac{x}{2} = 11}$

$\dfrac{x}{2} \cdot 2 = 11 \cdot 2$ EPO–<u>M</u>

$\boxed{x = 22}$

2. (a)

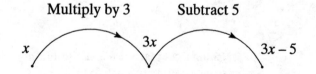

Multiply by 3   Subtract 5

x → $3x$ → $3x - 5$

(b)

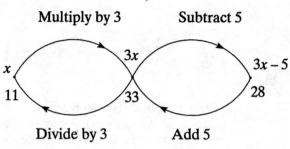

Multiply by 3   Subtract 5

$3x$ → $3x - 5$

x 11   33   28

Divide by 3   Add 5

(c) $\boxed{3x - 5 = 28}$

$3x - 5 + 5 = 28 + 5$ EPO–<u>A</u>

$\boxed{3x = 33}$

$3x \div 3 = 33 \div 3$ EPO–<u>D</u>

$\boxed{x = 11}$

3. (a)

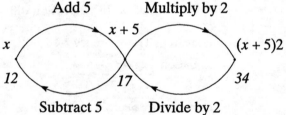

Add 5   Multiply by 2

x → $x + 5$ → $(x + 5)2$

(b)

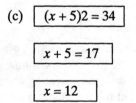

Add 5   Multiply by 2

$x + 5$ → $(x + 5)2$

x 12   17   34

Subtract 5   Divide by 2

(c) $\boxed{(x + 5)2 = 34}$

$\boxed{x + 5 = 17}$

$\boxed{x = 12}$

(d) 12 is a solution because substituting 12 for x makes a true statement. $(12 + 5)2 = 34.$

4. (a)

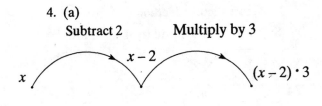

(b)

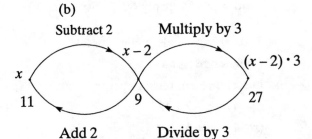

(c) $(x-2)3 = 27$

$x - 2 = 9$

$x = 11$

(d) 11 is a solution of $(x-2) \cdot 3 = 27$ because substituting 11 for x makes a true statement. $(11-2) \cdot 3 = 27$.

5. Note: Encourage students who can do so to write the series of equations without drawing the arrow pictures. Encourage students who still need the arrow pictures to use them.

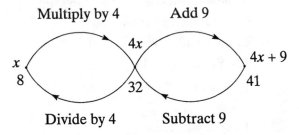

$4x + 9 = 41$

$4x = 32$

$x = 8$

Note: $4(8) + 9 = 41$ is a true statement.

Lesson 2-8: Problem Solving with Arrow Pictures and Equations

Objective

To solve problems using arrow pictures and equations

Prerequisite Lessons

Lessons 2-1, 2-2, 2-3, 2-4, 2-5, 2-6, and 2-7

Materials Needed

For each student: Worksheets 2-8-1, 2-8-2
Transparencies 2-8-1, 2-8-2

Directions for the Teacher

If students are not aware of the standard agreement on order of operations, it may be necessary to investigate and/or review this agreement.

Place the first transparency on an overhead projector and read the problem presented there. With students, complete an arrow picture solution. Such a solution follows:

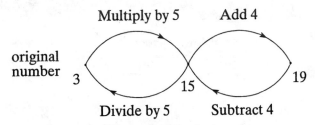

Thus, the original number was 3.

Next, lead the students through an equation solution method. Such a solution follows:

$5x + 4 = 19$

$5x = 15$

$x = 3$

Thus, the original number was 3.

Place the second transparency on the overhead projector and proceed to an arrow picture solution similar to the one that follows:

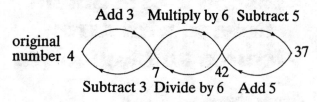

Thus, the original number was 4.

Proceed to the equation solution method. Emphasize the need for parentheses around $x + 3$ because of the order of operation agreement. A solution follows.

$$(x + 3)6 - 5 = 37$$
$$(x + 3)6 = 42$$
$$x + 3 = 7$$
$$x = 4$$

Thus, the original number was 4.

Note: Some calculators observe the order of operations. A student using such a calculator will need to press = after each new operation to force the calculator to add or subtract before it multiplies or divides.

Distribute worksheets. Students may find problems 2 and 3 more difficult. Provide individual assistance as needed.

Answer Key

1. **Arrow Picture Method**

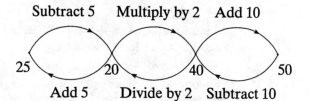

Thus, the original number was 30.

Equation Method

Let x represent the number.

$$x \div 3 - 4 = 6$$
$$x \div 3 = 10$$
$$x = 30$$

Thus, the original number was 30.

2. **Arrow Picture Method**

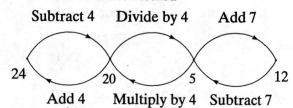

Thus, the original number was 24.

Equation Method

Let x represent the number.

$$(x - 4) \div 4 + 7 = 12$$
$$(x - 4) \div 4 = 5$$
$$x - 4 = 20$$
$$x = 24$$

Thus, the original number was 24.

3. **Arrow Picture Method**

Subtract 5   Multiply by 2   Add 10

25   20   40   50

Add 5   Divide by 2   Subtract 10

Her allowance was $25.

Equation Method

Let x represent her allowance.

$$(x - 5)2 + 10 = 50$$
$$(x - 5)2 = 40$$
$$x - 5 = 20$$
$$x = 25$$

Her allowance was $25.

Lesson 2-9: Applications in Problem Solving

Objective

To express numerical relationships in word problems as equations.

To use algebraic representations to find reasonable solutions to word problems.

Prerequisite Lessons

Lessons 2-1, 2-2, 2-3, 2-4, 2-5, 2-6, 2-7, and 2-8

Materials Needed

For each student: Worksheet 2-9
Transparency 2-9

Directions for the Teacher

Place the transparency on an overhead projector. Have students work in small groups to solve the first problem is *any* way they can. Ask students to share their solutions.

After students have established that it costs 50¢ to play once, discuss how to translate their solutions into equations. There are two unknown amounts: Maria's allowance and the cost of playing once. The latter is what is asked for in the problem. Choose a variable to represent the cost to play once. Write what the variable represents: $p \rightarrow$ cost in cents of playing once. (The arrow indicates what p represents.)

Known relationships: $6p + 300 \rightarrow$ Allowance

$10p + 100 \rightarrow$ Allowance

We know that $6p + 300 = 10p + 100$. Relate this to a balance scale problem. Write equations that would result from equality-preserving operations.

$6p + 300 = 10p + 100$

$300 = 4p + 100$

$200 = 4p$

$50 = p$

Have students proceed to the second problem. After students have solved the problem in *any* way, discuss their solutions. If someone used an arrow picture, capitalize on that by linking it to the equation to represent the problem (maybe covering arrow picture).

Desired information: m $\rightarrow$ original amount in dollars

Known relationships: m $-$ 3.00 $\rightarrow$ amount after hot dog and cola

(m $-$ 3.00) $\div$ 2 $\rightarrow$ amount after rides

(m $-$ 3.00) $\div$ 2 $-$ 1.50 $\rightarrow$ amount after cotton candy

Ask students to record the equations that would result from equality-preserving operations.

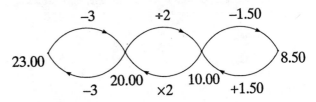

$(m - 3) \div 2 - 1.50 = 8.50$

$(m - 3) \div 2 = 10.00$

$m - 3 = 20.00$

$m = 23.00$ (Note that it is important to remember what the variable represents.)

Thus, he started with $23.00

Distribute a copy of the worksheet to each student to work independently or cooperatively.

Encourage them to represent and solve each problem algebraically.

Answer Key

1. $x \rightarrow$ Lynn's money in dollars

   $3x \rightarrow$ Jo's money in dollars

   $x + 3x = 36$

   $4x = 36$

   $x = 9$

   Lynn has $9. Jo has $27.

2. $x \rightarrow$ 6:00 A.M. temperature

   $2x + 5 = 39$

   $2x = 34$

   $x = 17$

   The temperature at 6:00 A.M. was 17°F.

3. $x \rightarrow$ number of hours worked

   $25 + 20x = 145$

   $20x = 120$

   $x = 6$

   The plumber worked for 6 hours.

4. $x \rightarrow$ number of tokens for 1 roller coaster ride

   $5x + 16 = 8x + 7$

$16 = 3x + 7$

$9 = 3x$

$3 = x$

It costs 3 tokens to ride the roller coaster.

When we write $2 + 1 = 3$, we use "=" to indicate that the quantity $2 + 1$ <u>is</u> <u>the</u> <u>same</u> <u>as</u> the quantity 3, and $2 + 1 = 3$ tells us how $2 + 1$ and 3 are related.

$2 + 1 = 3$ is a true statement.

$2 + 1 = 4$ is a false statement.

1. Is $3 + 5 = 4 + 4$ a true statement? Why?

2. Is $3 \cdot 5 = 10 \div 2 + 8$ a true statement? Why?

3. Is $x + 3 = 5$ a true statement? Why?

||||||||||||||||||||||||||||||||||||||||||||| **Equality** |||||||||||||||||||||||||||||||||||||||||||||

1. Which of the following equality statements are true?

   (a) $4 \cdot 3 = 14 - 2$

   (b) $4 \cdot 3 = 12 + 6$

   (c) $36 \div 4 = 14 - 5$

   (d) $5 - 2 = 18 \div 9$

2. Fill in the blank to make each sentence true.

   (a) $7 \cdot 4 = 23 + $ _____

   (b) _____ $\cdot 6 = 25 - 7$

   (c) _____ $\div 3 = 8 - 4$

   (d) $2 + 2 = 36 \div $ _____

The number 4 is a solution of the equation $x - 1 = 3$, because when we substitute 4 for x, a true sentence is the result. More specifically, 4 is a solution of the equation $x - 1 = 3$ because
$$4 - 1 = 3$$
is a true sentence.

Notice that 5 is not a solution of this equation because
$$5 - 1 = 3$$
is a false sentence.

1. Which of the numbers 1, 2, 3, and 4 are a solution of the equation $2x - 1 = 7$?

2. Which of the numbers 4, 5, 6, and 7 are solutions of the equation $(x - 2)(x + 3) = 36$?

3. Write an equation which has $x = 5$ as a solution.

▮▮▮▮▮▮▮▮▮▮▮▮▮▮ **Introduction to Equations** ▮▮▮▮▮▮▮▮▮▮▮▮▮▮▮

1. Which of the numbers 5, 6, 7, and 8 are solutions of the equation $x + 7 = 13$?

2. Which of the numbers 7, 8, 9, and 10 are solutions of the equation $3x - 2 = 19$?

3. Which of the numbers 2, 3, 4, and 5 are solutions of the equations $\frac{x-1}{x+5} = \frac{1}{4}$?

4. Which of the numbers 1, 2, 3, 4, and 5 are solutions of the equation $(x + 2)(x + 3)(x + 7) = 180$?

5. Which of the numbers 1, 2, 3, 4, and 5 are solutions of the equations $\frac{1}{x} + \frac{1}{2x} = \frac{1}{2}$?

6. Which of the numbers 2, 4, 6, 8, and 10 are solutions to the equation $(x + 5)(x - 6) = 0$?

7. Write an equation that has $x = 4$ as a solution.

8. Write an equation that has $x = 2$ as a solution.

9. Write an equation that has both $x = 1$ and $x = 3$ as solutions.

||||||||||||||||||||||| **Introduction to Equations** |||||||||||||||||||||||

10. Which of the numbers $-2, -3, -4,$ and -5 are solutions of the equation $-2x + 5 = 15$?

11. Which of the numbers $-5, -6, -7,$ and -8 are solutions of the equation $2x + 23 = 11$?

12. Which of the numbers $-5, -6, -7,$ and -8 are solutions of the equation $(x + 7)(x + 5) = 0$?

Label the dots.

Multiply by 2   **Subtract 3**

1.

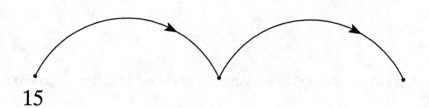

15

Subtract 10   **Multiply by 3**   **Divide by 2**

2.

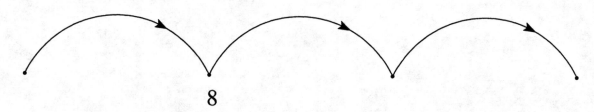

8

               50 Pre-Algebra Activities

1. In each case, the arrow means add 3.

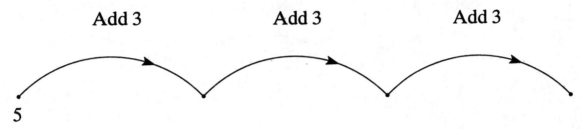

Add 3   Add 3   Add 3

5

    (a) Find a number corresponding to each unlabeled dot.

    (b) Draw in all "reverse" arrows.

    (c) What could these reverse arrows mean?

2. In each case, the arrow means multiply by 2.

    Multiply by 2   Multiply by 2   Multiply by 2

3

    (a) Find a number corresponding to each unlabelled dot.

    (b) Draw in all the reverse arrows.

    (c) What could these reverse arrows mean?

IIIIIIIIIII **Arrow Pictures and Inverse Operations** IIIIIIIIIII

1. In each case, the arrows mean subtract 10.

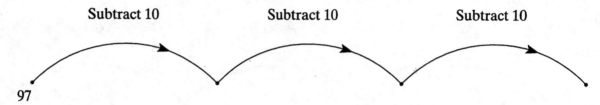

Subtract 10   Subtract 10   Subtract 10

97

    (a) Find a number corresponding to each unlabeled dot.

    (b) Draw in all reverse arrows.

    (c) What could the reverse arrows mean?

2. In each case the arrow means divide by 3.

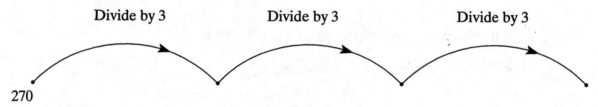

Divide by 3   Divide by 3   Divide by 3

270

    (a) Find a number corresponding to each unlabeled dot.

    (b) Draw in all the reverse arrows.

    (c) What could the reverse arrows mean?

׀׀׀׀׀׀׀׀׀׀ **Arrow Pictures and Inverse Operations** ׀׀׀׀׀׀׀׀׀׀

3. Study problems 1 and 2 and look for patterns of regular arrows and reverse arrows. Then complete the table below.

Regular Arrow	Reverse Arrow
Add 2	
Subtract 5	
Divide by 7	
Multiply by 100	
	Divide by 3
	Add 7

A physical situation related to equality is a balanced scale. When the scale balances, the weight on the left side equals the weight on the right side. Scale 1 balances.

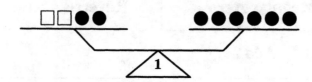

Figures that look alike weigh the same amount. Figures that look different have different weights. The following scales use the same weights as scale 1. Explain why each one balances.

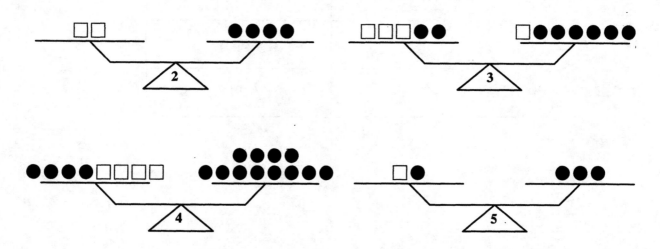

The balance-preserving operations for scales relate to equality-preserving operations for equations.

Balance-Preserving Operations

Equality-Preserving Operations

$6 = 6$

BPO-A. Add the same weight to each side of the scale.

EPO-A. Add the same number to each side of the equation.

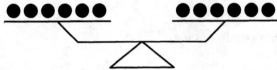

$6 + 2 = 6 + 2$

BPO-S. Subtract the same weight from each side of the scale.

EPO-S. Subtract the same number from each side of the equation.

$6 - 2 = 6 - 2$

BPO-M. Multiply the weight on each side of the scale by the same factor.

EPO-M. Multiply each side of the equation by the same number.

$6 \cdot 3 = 6 \cdot 3$

BPO-D. Divide the weight on each side of the scale by the same divisor.

EPO-D. Divide each side of the equation by the same number.

$6 \div 3 = 6 \div 3$

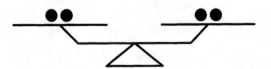

50 Pre-Algebra Activities

||||||||||||||| **Balance-Preserving Operations** |||||||||||||||

1. The scale shown here balances. Objects that look the same weigh the same.

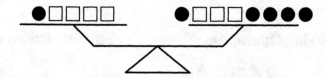

Draw two more arrangements of the weights that would balance.

(a)

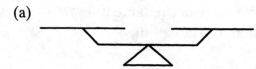

Explain why this scale balances: _____

_____

_____

(b)

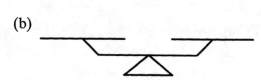

Explain why this scale balances: _____

_____

_____

_____

(c) Without changing the weight on the right side, what weights could be put on the left side of this scale to make it balance?

Explain why this scale balances: _____

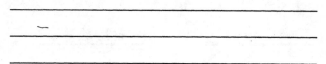

_____

_____

_____

Name _____

Date _____

IIIIIIIIIIIIIIIIIIIII **Balance-Preserving Operations** IIIIIIIIIIIIIIIIIIIII

2. The scale shown here balances. Objects that look the same have the same weight.

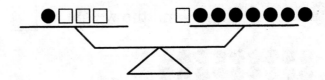

Draw two more arrangements of the weights that would balance.

(a) 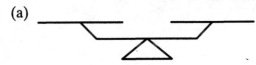   Explain why this scale balances: _____

_____

_____

_____

(b) 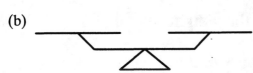   Explain why this scale balances: _____

_____

_____

_____

(c) Without changing the weight on the left side, what weights could be put on the right side of this scale to make it balance?

Explain why this scale balances: _____

_____

_____

_____

                   50 Pre-Algebra Activities

1. Suppose ● weighs 1 gram. Find the weight of ☐.

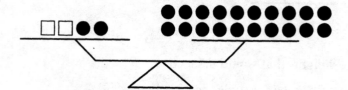

Explanation:

2. Suppose ● weighs 1 gram. Find the weight of ☐.

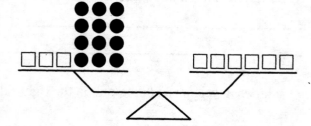

Explanation:

50 Pre-Algebra Activities

||||||||||||||||| **Solving Weight Balance Problems** |||||||||||||||||

In each problem, assume that the scale balances and that ● weighs 1 gram. Tell which balance-preserving operations you use in each problem.

1. Find the weight of ☐ if the scale shown below balances.

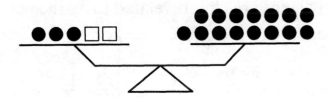

2. Find the weight of ☐.

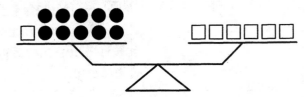

3. Find the weight of ☐.

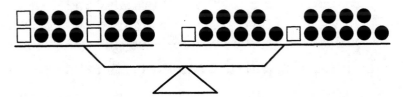

4. Find the weight of ☐.

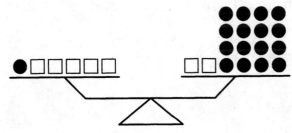

Mathematicians use variables to represent unknown quantities in equations. A balanced scale can be represented by an equation.

Operations that preserve balance are closely related to mathematical operations that preserve an equality relationship.

Suppose ● weighs 1 gram. How can we find the weight of □?

Scale	Related Equation	Reason
	$x + 3 = 8$	x represents the weight of □ on the left side. There are 3 more grams on the left. There are 8 grams on the right.
	$x = 5$	Keeping the scales balanced, we removed 3 grams from each side of the scales. So □ weighs 5 grams.

Notice that when we substitute 5 for x in the equations, the statements are true.

$$\underline{5} + 3 = 8$$

5 is the solution because it makes each equation true.

Suppose ● weighs 1 gram. Write an equation that corresponds to the balance scale shown here. What is the weight of □?

Scale	Equation	Reason
		The twelve grams can be divided into 3 groups with 4 grams in each group. If we divide the number of □ by 3, we get 1 □. If we divide 12 by 3, we get 4. When we divide the weight on each side by 3, we preserve the balance.

Check your solution by substituting 4 for *x* in the original equation. Is the resulting statement true? If you substitute 4 ● for each □ on the first scale, will it balance?

50 Pre-Algebra Activities

Suppose ● weighs 1 gram. What is the weight of □? Write an equation that represents each balanced scale. Explain how the equation represents the balanced scale. When you make changes on the scales, explain how your changes keep the scales balanced. Check your solution both with scales and by substituting in each equation.

Equation:

EPO–____   _____

Equation:

EPO–____   _____

Equation:

Linking Scales and Equations

Suppose ● weighs 1 gram. Write an equation that represents the balanced scale here. Explain your equation. Find the weight of □, giving the scales, equations, and reasons for your solution. Be sure to check your solution with the original situation.

1.

Equation: _____

EPO–___ _____

Equation: _____

2.

Equation: _____

EPO–___ _____

Equation: _____

50 Pre-Algebra Activities

ⅢⅢⅢⅢⅢⅢⅢⅢ **Linking Scales and Equations** ⅢⅢⅢⅢⅢⅢⅢⅢ

3.

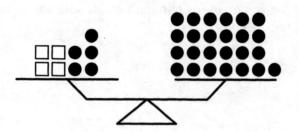

Equation: [   ]

EPO–____   _____

Equation: [   ]

EPO–____   _____

Equation: [   ]

4.

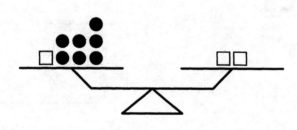

Equation: [   ]

EPO–____   _____

Equation: [   ]

Name _____

Date _____

IIIIIIIIIIIIIIIIIII **Linking Scales and Equations** IIIIIIIIIIIIIIIIIII

5.

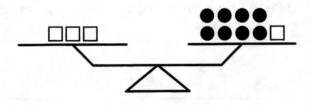

Equation: _____

EPO–___ _____

Equation: _____

EPO–___ _____

Equation: _____

6.

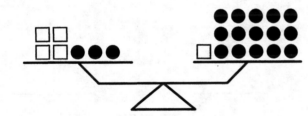

Equation: _____

EPO–___ _____

Equation: _____

EPO–___ _____

Equation: _____

EPO–___ _____

Equation: _____

50 Pre-Algebra Activities

Name _____

Date _____

IIIIIIIIIIIIIIIII Linking Scales and Equations IIIIIIIIIIIIIIIII

Suppose ● represents 1 gram and the weight of □ is unknown. Draw a balanced scale to represent each equation. Write the equations that would result from any balance-preserving operation to find the weight of □.

1. $2x + 7 = 13$

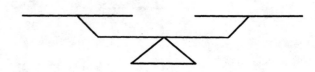

2. $3x + 8 = 2x + 14$

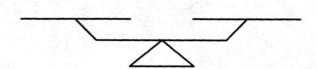

50 Pre-Algebra Activities

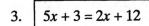

Linking Scales and Equations

3. | $5x + 3 = 2x + 12$ |

1. (a) Label the dots with appropriate expressions involving x.

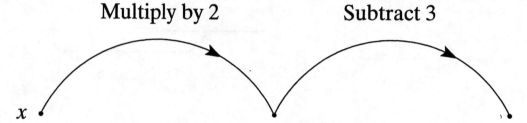

Multiply by 2   Subtract 3

x

(b) If $x = 6$, then $2x - 3 =$ _____.

If $x = 14$, then $2x - 3 =$ _____.

(c) If $2x - 3 = 15$, then $x =$ _____.

(d) One equation suggested by the arrow picture is
$2x - 3 = 15$. Write 2 other equations involving x suggested
by the arrow picture. Tell what equality-preserving operations were performed to make the other equations.

Equation: | $2x - 3 = 15$ |

_____   EPO–_____

Equation: | |

_____   EPO–_____

Equation: | |

2. (a) Label the arrows with the appropriate operations.

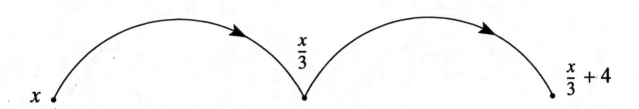

$$x \qquad \frac{x}{3} \qquad \frac{x}{3} + 4$$

(b) If $\frac{x}{3} + 4 = 6$, then $x = $ _____.

(c) Write the 3 equations suggested by the arrow picture. Tell what equality-preserving operations were performed.

Equation: [   ]

_____   EPO– _____

Equation: [   ]

_____   EPO– _____

Equation: [   ]

ꜛꜛꜛꜛꜛꜛꜛꜛ Solving Equations with Arrow Pictures ꜛꜛꜛꜛꜛꜛꜛꜛ

1. (a) Label the dots with appropriate expressions involving x.

Divide by 2   Add 6

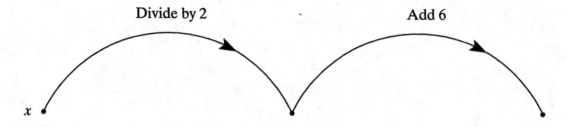

x

(b) Use the arrow picture in part (a) to solve the equation $\frac{x}{2} + 6 = 17$.

(c) Write the other equations suggested by the arrow picture. Tell what equality-preserving operations were performed.

Equation: $\boxed{\dfrac{x}{2} + 6 = 17}$

_____   EPO–_____

Equation: [   ]

_____   EPO–_____

Equation: [   ]

50 Pre-Algebra Activities

ⅢⅢⅢⅢ **Solving Equations with Arrow Pictures** ⅢⅢⅢⅢⅢ

2. (a) Label the arrows on the arrow picture shown here.

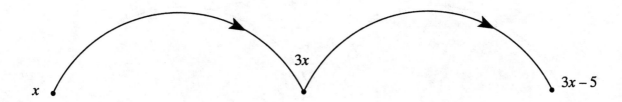

x $3x$ $3x - 5$

(b) Use the arrow picture to solve $3x - 5 = 28$.

(c) Write the equations suggested by the arrow picture. Tell what equality-preserving operations were performed.

Equation: _____

_____   EPO–_____

Equation: _____

_____   EPO–_____

Equation: _____

ıııııııııı Solving Equations with Arrow Pictures ıııııııııı

3. (a) Label the arrows on the arrow picture shown here.

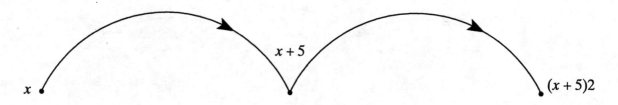

(b) Use the arrow picture to solve $(x + 5) \cdot 2 = 34$

(c) Write the equations suggested by the arrow picture.

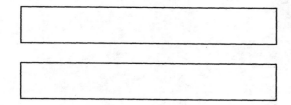

(d) Why is this value of x a solution of $(x + 5) \cdot 2 = 34$?

ıııııııııı Solving Equations with Arrow Pictures ıııııııı

4. (a) Label the second dot with an expression involving x. Label the arrows with appropriate operations.

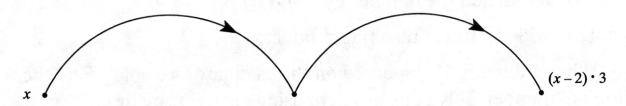

$$x \qquad\qquad\qquad\qquad\qquad\qquad (x-2) \cdot 3$$

   (b) Use the arrow picture to solve $(x-2)3 = 27$.

   (c) Write the equations suggested by the arrow picture.

   ┌─────────────────────────────┐
   │   │
   └─────────────────────────────┘

   ┌─────────────────────────────┐
   │   │
   └─────────────────────────────┘

   ┌─────────────────────────────┐
   │   │
   └─────────────────────────────┘

   (d) Why is this value of x a solution of $(x-2)3 = 27$?

5. Solve $4x + 9 = 41$

50 Pre-Algebra Activities

I entered a number in my calculator. Then I

  (i) multiplied the number by 5 and

 (ii) added 4 to the resulting product.

When I did this, 19 appeared on the calculator's display. Find the original number. Solve this problem using an arrow picture. Then solve it by finding an equation and solving the equation.

Arrow Picture Method

_____

Equation Method

I entered a number in my calculator. Then I

(i)   added 3 to the number,

(ii)   multiplied the sum by 6, and

(iii)   subtracted 5 from the product.

When I did this, 37 appeared on the calculator's display. Find the original number. Solve this problem using an arrow picture. Then solve it by finding an equation and solving the equation.

Arrow Picture Method

Equation Method

Solving Problems with Arrow Pictures and Equations

‖‖‖‖‖‖‖‖‖‖‖‖‖‖‖‖‖‖‖   ‖‖‖‖‖‖‖‖‖‖‖‖‖‖‖‖‖‖‖

For each of the problems below, solve the problem using an arrow picture. Then solve it by finding an equation and solving the equation.

1. I entered a number in my calculator. Then I

   (i) divided the number by 3 and

   (ii) subtracted 4 from the resulting number.

   When I did this, 6 appeared on the calculator's display. What was the original number?

   Arrow Picture Method

   _____

   Equation Method

2. I entered a number in my calculator. Then I

   (i) subtracted 4 from the number,

   (ii) divided the resulting number by 4, and

   (iii) added 7 to that number.

   When I did this, 12 appeared on the calculator's display. What was the original number?

   Arrow Picture Method

   _____

   Equation Method

Solving Problems with Arrow Pictures and Equations

|||||||||||||||||||||| ||||||||||||||||||||||

3. Mary was given her allowance. She went to a movie and spent $5 of her allowance. Then she mowed the neighbor's lawn and was paid for doing this. As a result, she now had twice as much money as she had after buying the movie ticket. When she received a birthday gift of a ten-dollar bill, she had a total of $50. What was the amount of her allowance?

Arrow Picture Method

_____

Equation Method

1. Maria took her allowance to the arcade. She can either play her favorite game 6 times and have $3.00 left over or she can play her favorite game 10 times and have $1.00 left. How much does it cost to play her favorite game once?

   Desired information:

   Known relationships:

2. Gerald went to the carnival with some money. After he got to the carnival, he bought a hot dog and soft drink for $3.00. He spent half of the rest of his money on rides. Then he bought some cotton candy for $1.50 for his sister. When he counted his remaining money, he had $8.50 left. How much money did he have when he arrived at the carnival?

ⅢⅢⅢⅢⅢⅢ **Applications in Problem Solving** ⅢⅢⅢⅢⅢⅢ

1. Jo has three times as much money as Lynn. Together they have $36. How much money does each person have?

2. At 6:00 A.M. Juan recorded the outside temperature. By noon, the temperature had doubled. At 3:00 P.M. it was 39°F, which was 5 degrees warmer than it had been at noon. What was the temperature at 6:00 A.M.?

3. A plumber charges $25 for traveling to your house and then charges $20 for each hour of work. If the bill was $145, how long did the plumber work?

4. Bess has a pocket full of tokens. If she rides the roller coaster 5 times, she will have 16 tokens left. If she rides it 8 times she will have 7 tokens left. How many tokens does it take to ride the roller coaster once?

Chapter 3

Graphing

In this chapter, students learn about two-dimensional graphs. In Lessons 3-1, 3-2, and 3-3, they learn about the two-dimensional coordinate system. In Lesson 3-4, they find equations of pictured lines. In Lesson 3-5, they actually graph lines, and in Lesson 3-6, they graph nonlinear figures including parabolas and circles. The last four lessons involve negative integers.

Lesson 3-1: Points and Ordered Pairs I

Objective

To identify ordered pairs for first quadrant points

Prerequisite Lessons

None

Materials Needed

For each student: Worksheet 3-1
Transparency 3-1

Directions for the Teacher

Place the transparency on an overhead projector. Emphasize that the first coordinate of an ordered pair indicates the number of units to the right and the second coordinate indicates the number of units up. With students, complete the table as follows.

Points	Ordered Pairs
A	(3,1)
B	(1,3)
C	(5,5)
D	(4,0)
E	(0,2)
F	(4,2)
G	(4,5)

Distribute worksheet and provide individual assistance as needed.

Answer Key

1.

Point	Ordered Pair
A	(5,3)
B	(3,4)
C	(1,1)
D	(0,4)
E	(2,5)
F	(4,2)

2.

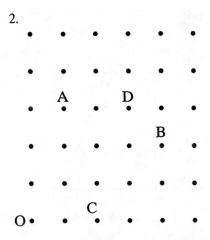

Lesson 3-2: Ordered Pairs for Points on Segments

Objective

To identify patterns for ordered pairs for points on segments.

Prerequisite Lessons

Lesson 3-1

Materials Needed

For each student: Worksheets 3-2-1-, 3-2-2
Transparency 3-2

Directions for the Teacher

Briefly review Lesson 3-1. Place the transparency on an overhead projector and proceed to problem 1. With appropriate student help, complete the table.

Point	Ordered Pair
A	(1,0)
B	(2,1)
C	(3,2)
D	(4,3)
E	(5,4)

Ask if they see any pattern involved with the ordered pairs. Proceed to problem 1b. They will prob-

ably suggest that the second component is 1 less than the first component or the first component is 1 greater than the second component. Go on to problem 1c. They will probably suggest that (6,5) and (7,6) are appropriate ordered pairs. Finally, direct their attention to problem 1d. They will probably conclude that one or more of the following equations are appropriate:

$$y = x - 1$$
$$x = y + 1$$
$$x - y = 1$$

If they suggest only one of these equations, encourage them to see if they can describe this relationship in other ways and, if necessary, lead them to the other equations.

Distribute worksheets and provide individual assistance as needed.

Answer Key

1. (a)

Point	A	B	C	D
Ordered Pair	(2,5)	(3,4)	(4,3)	(5,2)

(b) (1,6) and (6,1) (Other answers are possible.)

(c) $x + y = 7$ (Other possible answers are $y = 7 - x$ and $x = 7 - y$.)

2. (a)

Point	A	B	C	D
Ordered Pair	(0,2)	(1,3)	(2,4)	(3,5)

(b) (4,6) and (5,7) (Other answers are possible.)

(c) $y = x + 2$ (Other possible answers are $x = y - 2$ and $y - x = 2$.)

3. (a)

Point	A	B	C	D
Ordered Pair	(1,4)	(2,3)	(3,2)	(4,1)

(b) (0,5) and (5,0)

(c) $x + y = 5$ (Other possible answers are
$y = 5 - x$ and $x = 5 - y$.)

Lesson 3-3: Points and Ordered Pairs II

Objective

To identify ordered pairs for points in the standard coordinate system.

Prerequisite Lessons

Lessons 3-1 and 3-2

Materials Needed

For each student: Worksheets 3-3-1, 3-3-2
Transparency 3-3

Directions for the Teacher

For this lesson, assume that students are familiar with integers and can compute with integers (including negative integers). Place the transparency on an overhead projector and direct students' attention to the origin (labeled O). Indicate that the horizontal line is called the x-axis and is labeled X, and that the vertical line is called the y-axis and is labeled Y. Ask about the ordered pair for point A (4,3). Next, direct their attention to point B. Indicate that if you start at the origin, you must go one unit to the left and two units up so the ordered pair for B is (-1,2). Emphasize when the initial movement is to the left, the first coordinate is negative. Direct their attention to point C. Indicate that if you start at the origin you must go 4 units to the right and 1 unit down, so the ordered pair for C is (4,-1). Emphasize that when the final movement is downward, the second coordinate is negative. Proceed to points D and E. The coordinates for those points are (-3,-2) and (0,2).

Distribute worksheets and provide assistance as needed.

Answer Key

1.

Point	Ordered Pairs
A	(3,2)
B	(4,–1)
C	(–2,0)
D	(–3,–3)
E	(0,–3)
F	(–4,2)

2.

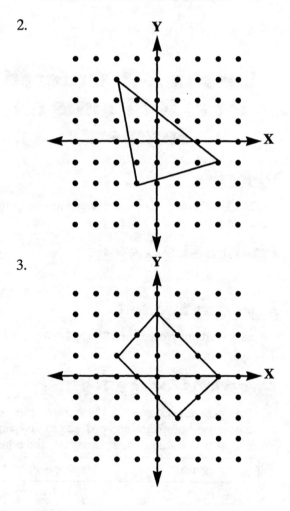

3.

Lesson 3-4: Finding Equations of Lines

Objective

Find ordered pairs and an equation for a pictured line

Prerequisite Lessons

Lessons 3-1, 3-2, and 3-3

Materials Needed

For each student: Worksheets 3-4-1, 3-4-2
Transparency 3-4

Directions for the Teacher

Briefly review Lesson 3-3. Place the transparency on the overhead projector and proceed to problem 1a. With students, select four points on the line and write the ordered pairs for these points. Proceed to problem 1b. Possible answers are $(-3,5)$ and $(5,-3)$. Proceed to problem 1c. Emphasize that you could call the first coordinate of an ordered pair x and the second component y. Ask students to look for patterns involving individual ordered pairs. Conclude with an equation. The most natural equation is

$$x + y = 2$$

but two other possibilities are

$$x = 2 - y$$

and

$$y = 2 - x$$

Distribute worksheets. Provide individual help as needed.

Answer Key

1. (a) $(-2,-2)$, $(-1,-1)$, $(0,0)$, $(3,3)$ (Others are possible.)

   (b) $(5,5)$, $(-6,-6)$ (Others are possible.)

   (c) $x = y$

2. (a) $(1,-2)$, $(2,-4)$, $(-1,2)$, $(-2,4)$, $(0,0)$

   (b) $(3,-6)$, $(-3,6)$ (Others are possible.)

   (c) $y = -2x$ (Another possibility is $x = -\frac{1}{2}y$.)

3. (a) $(-2,0)$, $(-1,-1)$, $(0,-2)$, $(1,-3)$ (Others are possible.)

   (b) $(3,-5)$, $(-5,3)$ (Others are possible.)

   (c) $x + y = -2$ (Others are possible.)

Lesson 3-5: Graphing Lines

Objective

To graph lines whose equations are given

Prerequisite Lessons

Lessons 3-1, 3-2, 3-3, and 3-4

Materials Needed

For each student: Worksheet 3-5
Transparency 3-5

Directions for the Teacher

Briefly review lesson 3-4. Tell the students that in lesson 3-4, they were given lines and asked to find equations of the lines, and in this lesson they will be given equations and asked to "graph" the lines.

Place the transparency on an overhead projector. With students, find four ordered pairs that meet the conditions of the equation $x + y = 4$. This can be done by selecting a value for x (or y) and calculating the corresponding value for y (or x). For instance, if x could be 3, then with students' help, conclude that the corresponding y value would be 1. Require that these x values be less than or equal to 4 because of limited possibilities of graphing points on the prepared grid. A possible table follows:

x	y	Order Pair
3	1	(3,1)
2	2	(2,2)
1	3	(1,3)
0	4	(0,4)

Plot these points on the graph, connect those points, and draw the corresponding line. Mention that all four of the points are actually on the line. Point out that it was only necessary to find two points but that the usual procedure is to find three points as a checking procedure.

Distribute worksheet and provide individual assistance as needed. Appropriate graphs follow.

Answer Key

1.

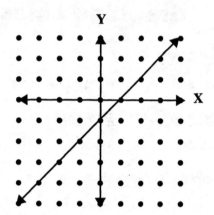

2.

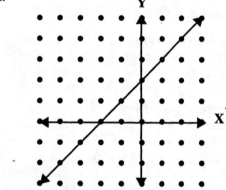

Lesson 3-6: Graphing Nonlinear Figures

Objective

To graph some nonlinear figures.

Prerequisite Lessons

Lessons 3-1, 3-2, 3-3, 3-4, and 3-5

Materials Needed

For each student: Worksheet 3-6-1

For each "advanced" student: Worksheets 3-6-2, 3-6-3

Transparency 3-6

Directions for the Teacher

Place the transparency on an overhead projector and direct the students' attention to the problem presented there. Suggest that the graphs of some equations are not lines and more specifically the graph of $x = y^2$ is not a line. Also suggest that values of y be chosen and then corresponding values of x can be found. Select y values of $-2, -1, 0, 1,$ and 2 and complete the table as shown below.

x	y	Ordered Pair
4	–2	(4,–2)
1	–1	(1,–1)
0	0	(0,0)
1	1	(1,1)
4	2	(4,2)

Point out that if you picked a value of 3 for y, then x would be 9 and that the pictured grid does not contain an x value greater than 5. Note that if you picked a value of -3 for y then x would be 9. Plot the points from the table and suggest that the graph would "continue" with the pattern suggested. Connect the related points as shown below.

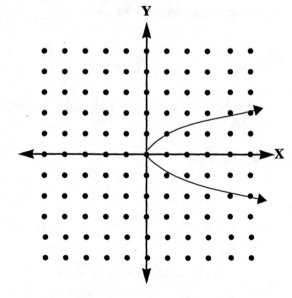

Distribute Worksheet 3-6-1. Have students proceed to problem 1. Provide individual help as needed. You may need to point out that if they chose *x* values greater than 2 or less than –2, the corresponding *y* values would be off the grid.

Answer Key

Solutions for Worksheet 3-6-1

1.

x	*y*	Ordered Pair
2	4	(2,4)
1	1	(1,1)
0	0	(0,0)
–1	1	(–1,1)
–2	4	(–2,4)

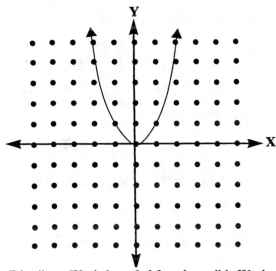

Distribute Worksheet 3-6-2 and possibly Worksheet 3-6-3 to advanced students.

2.

x	*y*	Ordered Pair
5	0	(5,0)
–5	0	(–5,0)
3	4	(3,4)
3	–4	(3,–4)
–3	4	(–3,4)
–3	–4	(–3,–4)
0	5	(0,5)
0	–5	(0,–5)
4	3	(4,3)
4	–3	(4,–3)
–4	3	(–4,3)
–4	–3	(–4,–3)

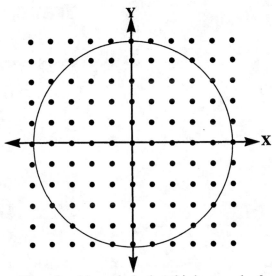

They should conclude that this is a graph of a circle. Individual students may need help with this problem.

A completed version for problem 3 follows. It is assumed that students already know about absolute value. Again, students may need individual help.

3.

x	*y*	Ordered Pair
0	0	(0,0)
1	1	(1,1)
–1	1	(–1,1)
2	2	(2,2)
–2	2	(–2,2)
3	3	(3,3)
–3	3	(–3,3)

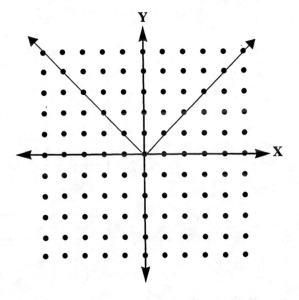

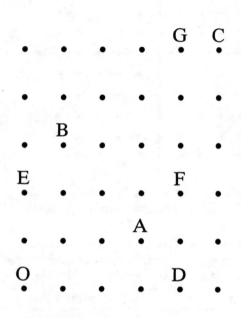

The point O is called the origin for this configuration of dots. The position of each of the other points is described in terms of its position relative to O. For example, to get from the origin to A you must go three units to the right and one unit up. This point is described by the ordered pair

$$(3, 1)$$

Three units   One unit up
to the right

In every case, the first component of the ordered pair describes the horizontal distance from the origin and the second component describes the vertical distance from the origin.

1. Complete the following table concerning points and the corresponding ordered pairs.

Points	Ordered Pairs
A	
B	
C	
D	
E	
	(4, 2)
	(4, 5)

50 Pre-Algebra Activities

xxxxxxxxxxx **Points and Ordered Pairs I** xxxxxxxxxxx

1. Complete the table concerning the points and ordered pairs.

Point	Ordered Pair
A	
B	
C	
D	
	(2, 5)
	(4, 2)

2. On the graph, label the points described in the table below.

Point	Ordered Pair
A	(1,3)
B	(4,2)
C	(2,0)
D	(3,3)

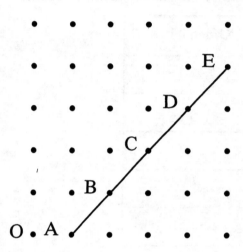

1. (a) Complete the table concerning the points on the pictured segment.

   (b) Describe (in words) how the first and second components of the ordered pairs are related.

Point	Ordered Pair
A	
B	
C	
D	
E	

   (c) Write the ordered pairs for two more points which would be on an extension of this segment.

   _____ , _____

   (d) The first component of an ordered pair is often referred to as *x* and the second component is often referred to as *y*. Write an equation which describes how *x* and *y* are related for an ordered pair (*x*,*y*) which is on the extension of the segment pictured.

×××××× **Ordered Pairs for Points on Segments** ××××××

1. (a)

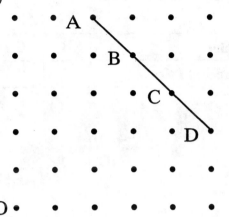

Complete the table below.

Point	A	B	C	D
Ordered Pair				

(b) List two ordered pairs for two points which would be on the extension of this segment.

(c) Write an equation which describes how the two components are related for points on this segment.

2. (a)

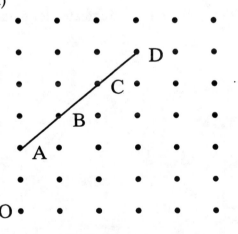

Complete the table below.

Point	A	B	C	D
Ordered Pair				

(b) List two ordered pairs for two points which would be on the extension of this segment.

(c) Write an equation which describes how the two components are related for points on this segment.

50 Pre-Algebra Activities

⟨×××××× **Ordered Pairs for Points on Segments** ××××××⟩

3. (a)

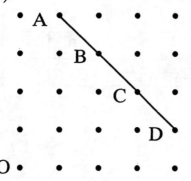

Complete the table below.

Point	A	B	C	D
Ordered Pair				

(b) List two ordered pairs for two points which would be on the extension of this segment.

(c) Write an equation which describes how the two components are related for points on this segment.

Complete the table below.

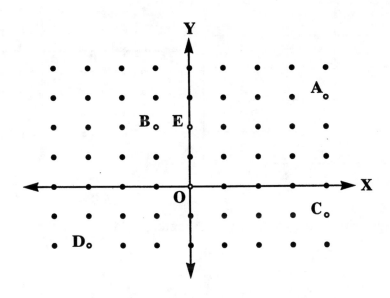

Point	A	B	C	D	E
Ordered Pair					

50 Pre-Algebra Activities

xxxxxxxxxxxxx**Points and Ordered Pairs II**xxxxxxxxxxx

1. Identify the ordered pairs for the points named.

Point	Ordered Pair
A	
B	
C	
D	
E	
F	

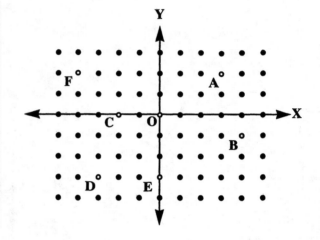

2. Draw a triangle with vertices at (3,–1), (–2,3) and (–1,–2).

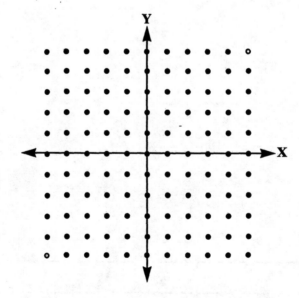

50 Pre-Algebra Activities

xxxxxxxxxxxx **Points and Ordered Pairs II** xxxxxxxxxx

3. Draw a rectangle with vertices at (–2,1), (1,–2), (3,0) and (0,3).

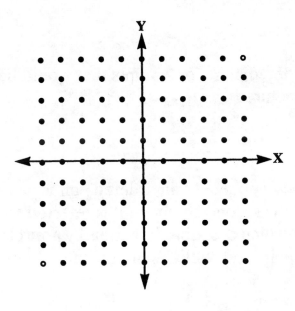

1. (a) For the line pictured below, write the ordered pairs for 4 points on the line (pictured).

   (b) Write ordered pairs for 2 points that would be on the line but are not pictured.

   (c) The first component of an ordered pair is generally referred to as *x,* and the second component is referred to as *y.* Write an equation which describes how *x* and *y* are related for an ordered pair (*x,y*), which is on the line.

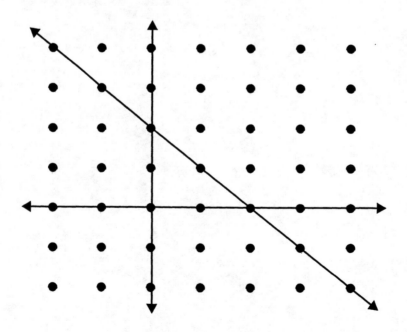

xxxxxxxxxxxxx **Finding Equations of Lines** xxxxxxxxxxx

1. (a) For the line pictured below, write the ordered pairs of 4 points pictured on the line.

   (b) For the line pictured below, write ordered pairs of 2 points that are on the line but are not pictured.

   (c) Write an equation of the line.

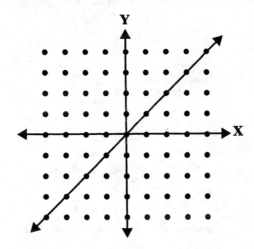

2. (a) For the pictured line below, write the ordered pairs of 5 points pictured on the line.

   (b) Write ordered pairs of 2 points that are on the line but are not pictured.

   (c) Write an equation of the line.

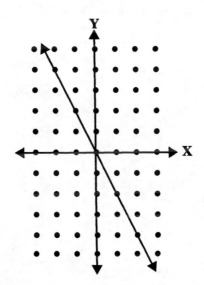

xxxxxxxxxxxx **Finding Equations of Lines** xxxxxxxxxxx

3. (a) For the line pictured below, write the ordered pairs of 4 points pictured on the line.

   (b) Write ordered pairs of 2 points that are on the line but are not pictured.

   (c) Write an equation of the line.

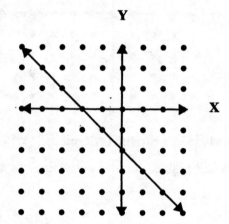

Graph the line whose equation is $x + y = 4$.

x	y	Ordered Pair

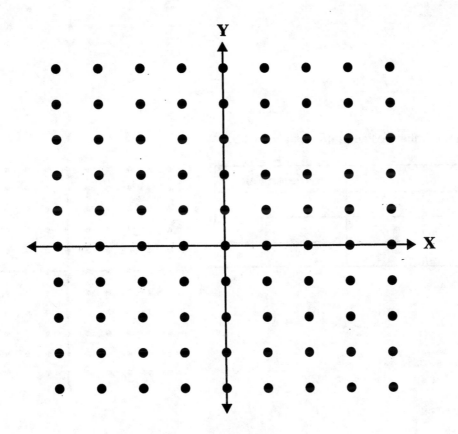

50 Pre-Algebra Activities

xxxxxxxxxxxxxxxxxx **Graphing Lines** xxxxxxxxxxxxxxxxxxxx

1. Graph the line whose equation is $x - y = 1$.

x	*y*	Ordered Pair

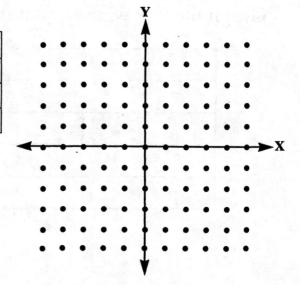

2. Graph the line whose equation is $y = x + 2$.

x	*y*	Ordered Pair

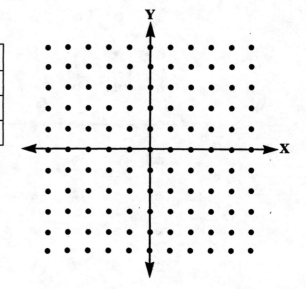

50 Pre-Algebra Activities

Graph the figure whose equation is $x = y^2$.

x	y	Ordered Pair

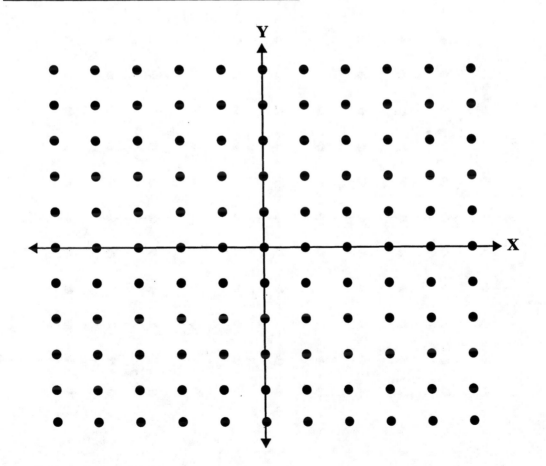

50 Pre-Algebra Activities

Name _____

Date _____

xxxxxxxxxxx **Graphing Nonlinear Figures** xxxxxxxxxxx

1. Graph the figure whose equation is $y = x^2$.

x	y	Ordered Pair

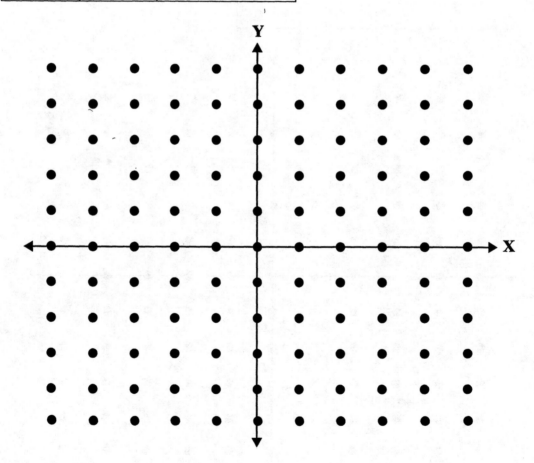

                   50 Pre-Algebra Activities

✕✕✕✕✕✕✕✕✕ **Graphing Nonlinear Figures** ✕✕✕✕✕✕✕✕✕

2. Graph the figure whose equation is $x^2 + y^2 = 25$.

x	y	Ordered Pair
5	0	(5,0)
–5		
3		
3		
–3		
–3		
0		
0		
4		
4		
–4		
–4		

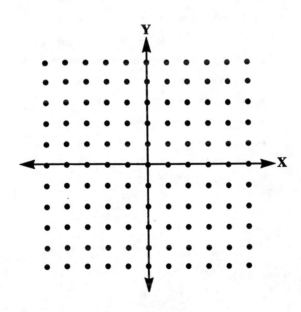

Name _____

Date _____

xxxxxxxxxx **Graphing Nonlinear Figures** xxxxxxxxxxx

3. Graph the figure whose equation is $y = | \, x \, |$.

x	y	Ordered Pair
0		
1		
−1		
2		
−2		
3		
−3		

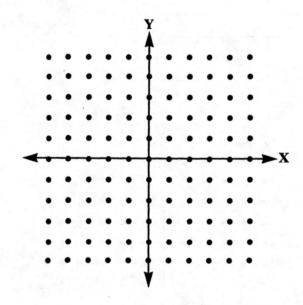

Chapter 4

Functions

If you were asked to give one word that describes mathematics, what would it be? A good answer to that question would be <u>functions</u>. This topic appears in a variety of circumstances. The binary operations of addition, subtraction, multiplication, and division are all functions.

The purpose of this chapter is to introduce students to functions. This is done by examining everyday situations and number relationships with respect to the following definition of function:

> A function is a correspondence between Set A
> and Set B in which each element of Set A
> "picks" exactly one element of Set B.

Then functions are presented by means of arrow pictures, tables, and equations.

Lesson 4-1: Introduction to Functions

Objective

To introduce the concept of function.

Prerequisite Lessons

Chapter 1

Materials Needed

For each student: Worksheets 4-1-1, 4-1-2
Transparencies 4-1-1, 4-1-2, 4-1-3, 4-1-4

Directions for the Teacher

Place the first transparency on the overhead projector. Introduce the idea of function in terms of a candy vending machine. Assume that the candy vending machine is full and the correct change has been inserted. In a properly working machine, pushing a button will release one type of candy. Note that it is acceptable for two different buttons to dispense the same kind of candy. The important idea is that a particular button picks the same kind of candy every time it is pushed.

Discuss how the candy vending machine illustrates the definition. Note what Set A and Set B are and how their elements are paired.

Place the second transparency on the overhead projector. Discuss why the two examples violate the definition of function. Note that in each case an element of the first set does not pick *exactly one* element of the second set. In that case, we might call the malfunctioning button a "fickle picker."

Place the third transparency on the overhead projector. Discuss the example. Students should refer to the definition of function to justify their responses.

This is a function because each total chooses exactly one shipping charge. It does not matter that two different total amounts would have the same shipping charge. That does not violate the definition of function.

Place the fourth transparency on the overhead projector. Fill in the chart by asking each student in the class his or her favorite ice cream flavor. Discuss the example. Students should refer to the definition of function to justify their responses. If someone does not have a favorite or if someone chooses two favorites, then the correspondence is not a function. Note, to be a function:

- Every person would have to choose a favorite flavor.
- More than one person may choose the same flavor.
- Some flavors may not be chosen by anyone.

Distribute worksheets. Students should read the example, determine whether the correspondence

described is a function, and justify their responses in terms of the definition of a function.

Answer Key

1.

Time/hrs	1	2	3	4	5	6	x
Distance/ Mi	60	120	180	240	300	360	$60x$

The correspondence is a function because each number of hours determines exactly one distance.

2.

Parking/ hrs	1	2	3	4	5	6	x
Parking Fee/$	5	7	9	11	13	15	$5 + 2(x-1)$ or $2x + 3$

The correspondence is a function because each number of hours determines exactly one parking fee.

3.

Weight/ oz.	1	2	3	4	5	6	x
Postage/ ¢	29	52	75	98	121	144	$29 + 23(x-1)$ or $23x + 6$

The correspondence is a function because each weight determines exactly one postage fee.

4.

Goals Attempted by Team A	1	2	3	4	5	6	x
Team A's Score	2 or 3	4, 5, or 6	6, 7, 8, or 9	8, 9, 10, 11, or 12	10, 11, 12, 13, 14, or 15	12, 13, 14, 15 16, 17, or 18	Values between and including $2x$ and $3x$

The correspondence is NOT a function because each number of goals could determine more than one score.

Lesson 4-2: Functions from Arrow Diagrams

Objective

To become aware of how functions can be shown with arrow diagrams.

Prerequisite Lessons

Lesson 4-1

Materials Needed

For each student: Worksheet 4-2
Transparencies 4-1-1, 4-2-1, 4-2-2

Directions for the Teacher

Place Transparency 4-1-1 from the previous lesson on an overhead projector. Tell students that the properly working candy machine can be represented in an arrow diagram. Place Transparency 4-2-1 on the overhead projector. Discuss how the arrow diagram represents a function. Ask students to locate the 2 sets. Ask students how they know each element of the first set picks exactly one element of the second set. They should conclude this is a function.

Place Transparency 4-2-2 on the overhead projector. Ask students to explain which arrow diagrams illustrate functions and which do not. Circle the appropriate label after each discussion. Correct answers follow:

1. Not a function. An element of the first set (John) does not choose exactly one element of the second set.

2. Function. Not all elements in the second set must be chosen.

3. Function. Elements of the first set can pick the same element of the second set. Some elements of the second set may not be chosen.

4. Not a function. An element of the first set picks more than one element of the second set.

Ask students to examine the two arrow diagrams that are functions and describe how they are alike. (They both have one arrow coming from every element in the first set.)

Ask students to describe how each of the non-examples are different from the function examples.

Distribute worksheet.

Answer Key

1. (a) No. 4 is a "fickle picker."

   (b) Yes.

   (c) No. 8 does not choose an element of Set B.

   (d) Yes.

   (e) No. 4 is a "fickle picker" and 2 does not choose an element of Set B.

   (f) Yes.

2. (Answers may vary. The diagram will represent a function only if there is exactly one arrow from each element of Set A to any element of Set B.)

Lesson 4-3: Functions from Tables

Objective

To determine what it means to have a table representing a function.

Prerequisite Lessons

Lessons 4-1 and 4-2

Materials Needed

For each student: Worksheet 4-3
Transparencies 4-2-1, 4-3-1, 4-3-2

Directions for the Teacher

Place Transparency 4-2-1 on an overhead projector. Review how the arrow diagram represents a function. Tell students another way to represent a function is through a table.

Place Transparency 4-3-1 on the overhead projector. Have students discuss how this table represents the same function as the arrow diagram.

Place Transparency 4-3-2 on the overhead projector. Discuss which tables represent functions. Correct answers follow.

(a) Not a function. 4 is a fickle picker.

(b) Not a function. 4 does not pick an element of B.

(c) Function. It is permissible for two elements of A to pick the same element of B.

(d) Function.

Distribute worksheets and provide assistance as needed.

Answer Key

1. (a) Not a function. (b) Function.

(c) Not a function.

2. (a) (Other answers are possible.)

A	B
2	9
4	7
6	5
8	3

(b) (Other answers are possible.)

A	B
2	9
4	7
6	5
8	3
8	5

Lesson 4-4: Functions from Equations

Objective

To determine a relationship between functions and equations.

Prerequisite Lessons

Lessons 4-1, 4-2, and 4-3

Materials Needed

For each student: Worksheets 4-4-1, 4-4-2
Transparency 4-4

Directions for the Teacher

Review the definition of a function. Suggest that the two sets need not be different sets. Tell students that for this lesson, all the functions will be from the set of all natural numbers to the set of all natural numbers. Place the transparency on the overhead projector and proceed to problem 1a. Indicate that it is common to call a general element from the first set (previously called A) x and to call a general element from the second set (previously called B) y. Have them investigate the relationship between x and y and describe what could be done to each element of set A to produce the corresponding element of Set B. Write this description as an equation ($y = x + 2$). Proceed to problem 1b. With students, conclude that the corresponding equation is $y = 2x$.

Distribute worksheets. Provide individual assistance as needed.

Answer Key

1. (a) $y = 2x + 1$

(b) $y = 3x$

x	y
1	2
2	6
3	10
4	14
•	•
•	•
•	•

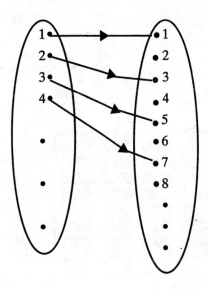

3. $y = x + 1$

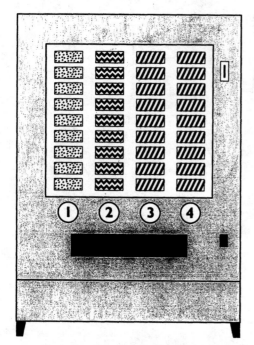

This machine has 4 buttons and 3 types of candy. Button 1 dispenses Candy X. Button 2 dispenses Candy Y. Button 3 and Button 4 dispense Candy Z, which is the most popular.

A properly working candy vending machine represents a function.

Set A: {button 1, button 2, button 3, button 4}

Set B: {Candy X, Candy Y, Candy Z}

Each button "picks" one type of candy to be dispensed.

A function is a correspondence between Set A and Set B in which each element of Set A "picks" exactly one element from Set B.

A function is a correspondence between Set A and Set B in which each element of Set A "picks" exactly one element from Set B.

If the machine is not working, pushing a button might release no candy. The correspondence would no longer be a function.

If the machine is not working, pushing a button might release one type of candy one time and a different type the next. The correspondence would no longer be a function.

Is the correspondence a function?

A catalog's shipping charge policy:

Total Price of Items	Shipping Charge
$0 to $30.00	$4.25
$30.01 to $70.00	$5.75
$70.01 or more	$6.95

Set A: All possible total price amounts

Set B: All possible shipping charges

Correspondence: Total price of purchase determines
a shipping charge

Is the correspondence a function?

Set A: people in this class

Set B: all flavors of ice cream

Relation: each person chooses a favorite flavor of ice cream

Set A	Set B

Set A	Set B

∿∿∿∿∿∿∿∿∿ Introduction to Functions ∿∿∿∿∿∿∿∿∿

1. Suppose that you are in a car travelling at a constant rate of 60 miles per hour. Let Set A be numbers of hours spent driving. Let Set B be the distances in miles. Consider a correspondence in which the number of hours determines the distance driven. Fill out the chart. Explain whether the correspondence is a function.

Time Spent Driving in Hours	1	2	3	4	5	6	x
Distance Travelled in Miles							

2. Suppose the rates for parking a car in a particular parking lot are $5 for the first hour and $2 for each additional hour. Let Set A be numbers of hours the car is parked. Let Set B be the parking fees. Consider a correspondence in which the number of hours parked determines the parking fee. Fill out the chart. Explain whether the correspondence is a function.

Hours the Car is Parked	1	2	3	4	5	6	x
Parking Fee in Dollars							

ᴧᴧᴧᴧᴧᴧᴧᴧᴧᴧ Introduction to Functions ᴧᴧᴧᴧᴧᴧᴧᴧᴧ

3. In past years the first class postal rates were 29¢ for the first ounce and 23¢ for each additional ounce. Let Set A be weights in ounces. Let Set B be the postage fees. Consider a correspondence in which the weight of the letter determines the postage fee. Fill out the chart. Explain whether the correspondence is a function.

Weight in Ounces	1	2	3	4	5	6	x
Postage Fees in Cents							

4. Consider a professional basketball game. Let Set A be possible numbers of field goal attempts made by Team A. Some field goals are worth 2 points and some are worth 3 points, depending on the distance. Let Set B be possible scores for Team A. Consider a correspondence in which the number of attempts taken determines the score. Fill out the chart. Explain whether the correspondence is a function.

Goals Made by Team A	1	2	3	4	5	6	x
Team A's Score							

Transparency 4-2-1

A function is a correspondence <u>between</u> Set A and Set B in which <u>each element of Set A "picks" exactly one element from Set B.</u>

1.

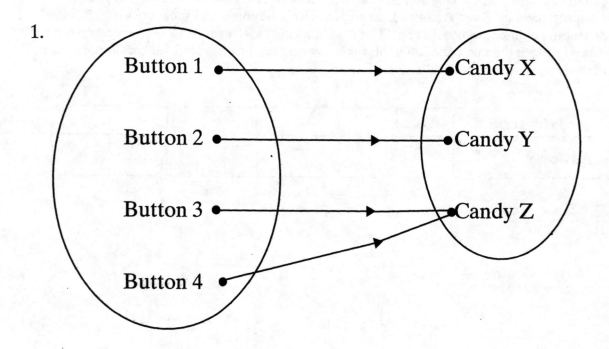

1.

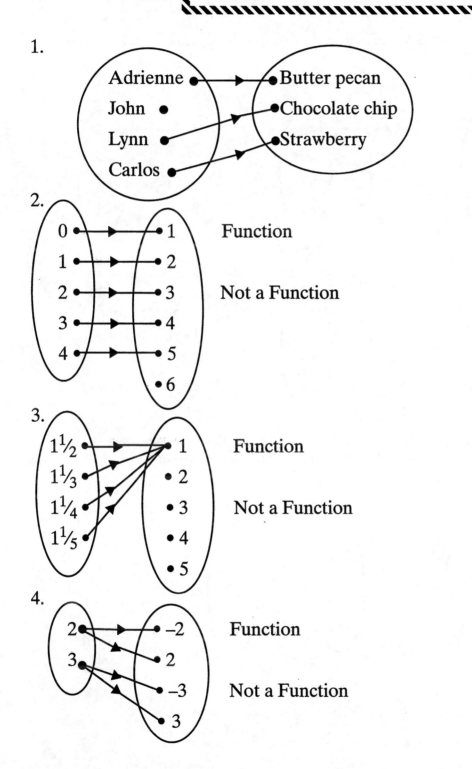

Adrienne → Butter pecan   Function

John ● Chocolate chip

Lynn ● Strawberry   Not a Function

Carlos ●

2.

0 → 1   Function

1 → 2

2 → 3   Not a Function

3 → 4

4 → 5

● 6

3.

$1\frac{1}{2}$ → 1   Function

$1\frac{1}{3}$ ● 2

$1\frac{1}{4}$ ● 3   Not a Function

$1\frac{1}{5}$ ● 4

● 5

4.

2 → −2   Function

3 ● 2

● −3   Not a Function

● 3

˄˄˄˄˄˄˄˄ Functions from Arrow Diagrams ˄˄˄˄˄˄˄˄

1. Given that A = {2,4,6,8} and B = {3,5,7,9}, which of the following arrow pictures describe functions from A to B?

a. A   B   b. A   B   c. A   B

 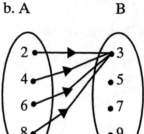 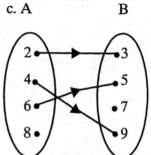

d. A   B   e. A   B   f. A   B

 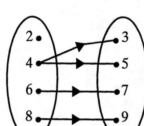 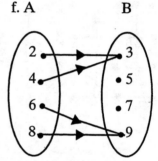

2. Given that A = {5,8,12,17} and B = {4,6,9,13}, draw an arrow picture which shows a correspondence between the elements of A and the elements of B that is

a. not a function.   b. a function.

A	B	A	B
5 •	• 4	5 •	• 4
8 •	• 6	8 •	• 6
12 •	• 9	12 •	• 9
17 •	• 13	17 •	• 13

50 Pre-Algebra Activities

A	B
Button 1	Candy X
Button 2	Candy Y
Button 3	Candy Z
Button 4	Candy Z

A function is a correspondence between Set A and Set B in which each element of Set A "picks" exactly one element from Set B.

A function is a correspondence between Set A and Set B in which each element of Set A "picks" exactly one element from Set B.

Given that A = {1,2,3,4} and B = {7,10,12,14,15}, which of the following tables describe functions from A to B?

a.

A	B
1	7
2	10
3	12
4	14
4	15

b.

A	B
1	7
2	10
3	15
4	

c.

A	B
1	7
2	7
3	14
4	14

d.

A	B
1	7
2	10
3	14
4	12

∿∿∿∿∿∿∿∿∿ Functions from Tables ∿∿∿∿∿∿∿∿∿

1. Given that A = {2,4,6,8} and B = {3,5,7,9}, which of the following tables describe a function from A to B?

(a)

A	B
2	3
4	5
4	7
6	9
8	9

(b)

A	B
2	3
4	3
6	3
8	3

(c)

A	B
2	3
4	9
6	5
8	

2. (a) Given that A = {2,4,6,8} and B = {3,5,7,9}, complete a table which describes a function from A to B.

A	B

(b) Complete a table which does *not* describe a function from A to B.

A	B

50 Pre-Algebra Activities

1. Write an equation for the function illustrated below.

   (a) Let *x* represent any element of Set A. Let *y* represent any element of Set B.

x	y
1	3
2	4
3	5
4	6
•	•
•	•
•	•

   (b) Let *x* represent any element of Set A. Let *y* represent any element of Set B.

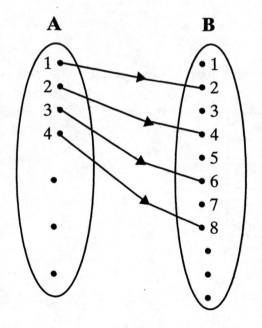

∧∧∧∧∧∧∧∧∧ **Functions from Equations** ∧∧∧∧∧∧∧∧∧∧∧

1. Write an equation for each function.

   (a)

x	y
1	3
2	5
3	7
4	9
•	•
•	•
•	•

   (b)

x	y
1	3
2	6
3	9
4	12
5	15
•	•
•	•
•	•

2. Complete a table for the function described by the equation $y = 4x - 2$.

x	y
1	
2	
3	
4	
•	•
•	•
•	•

50 Pre-Algebra Activities

⋀⋀⋀⋀⋀⋀ Functions from Equations ⋀⋀⋀⋀⋀⋀

3. Write an equation for the function pictured below.

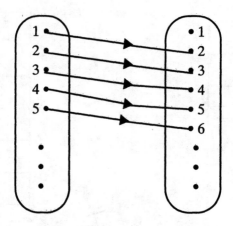

4. Draw an arrow picture for the function with $y = 2x - 1$ as the equation.

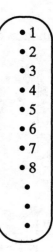

50 Pre-Algebra Activities